The protest business?

Issues in Environmental Politics

series editors Tim O'Riordan, Arild Underdal and Albert Weale

As the millennium approaches, the environment has come to stay as a central concern of global politics. This series takes key problems for environmental policy and examines the politics behind their cause and possible resolution. Accessible and eloquent, the books make available for a non-specialist readership some of the best research and most provocative thinking on humanity's relationship with the planet.

already published in the series

Environment and development in Latin America: the politics of sustainability *David Goodman and Michael Redclift (eds)*

The greening of British party politics *Mike Robinson*

The politics of radioactive waste disposal *Ray Kemp*

The new politics of pollution *Albert Weale*

Animals, politics and morality *Robert Garner*

Realism in Green politics *Helmut Wiesenthal (ed. John Ferris)*

Governance by green taxes: making pollution prevention pay *Mikael Skou Andersen*

Life on a modern planet: a manifesto for progress *Richard North*

The politics of global atmospheric change *Ian H. Rowlands*

Valuing the environment *Raino Malnes*

The making of European environmental policy *Duncan Liefferink*

forthcoming

European environmental policy: the pioneers:
   Mikael Skou Andersen and Duncan Liefferink (eds)

Sustaining Amazonia: grassroots action for productive conservation *Anthony Hall*

The protest business?

Mobilizing campaign groups

Grant Jordan and William A. Maloney

Manchester University Press
Manchester and New York
Distributed exclusively in the USA by St. Martin's Press

Published by Manchester University Press
Oxford Road, Manchester M13 9NR, UK
and Room 400, 175 Fifth Avenue, New York, NY 10010, USA

Distributed exclusively in the USA
by St. Martin's Press, Inc., 175 Fifth Avenue, New York, NY 10010, USA

British Library Cataloguing-in-Publication Data
A catalogue record for this book is available from the British Library.

Library of Congress Cataloging-in-Publication Data applied for

GE195
.J67
1997

ISBN 0 7190 4370 0 hardback
     0 7190 4371 9 paperback

First published in 1997

01 00 99 98 97   10 9 8 7 6 5 4 3 2 1

Typeset in Sabon
by Northern Phototypesetting Co. Ltd, Bolton
Printed in Great Britain
by Redwood Books, Trowbridge

Contents

List of figures and tables

Abbreviations

AA	Automobile Association
ASH	Action on Smoking and Health
APSA	American Political Science Association
CCI	Citizens for Community Improvement
CIWF	Compassion in World Farming
CND	Campaign for Nuclear Disarmament
CPRE	Council for the Protection of Rural England
CSO	Central Statistical Office
DMR	direct marketing recruit
DoE	Department of the Environment
ETA	Environmental Transport Association
FoE	Friends of the Earth
MSO	monthly standing order
NFU	National Farmers' Union
NOW	National Organization for Women
NSM	new social movement
PAC	Political Action Committee
PSMO	professional social movement organization
RM	resource mobilization
ROI	return on investment
RSPB	Royal Society for the Protection of Birds
RSPCA	Royal Society for the Prevention of Cruelty to Animals
SCCOPE	Sierra Club Committee on Political Education
SCLDF	Sierra Club Legal Defense Fund
SES	socio-economic status
SM	social movement
SMO	social movement organization
SNR	social network recruit
SSPCA	Scottish Society for the Prevention of Cruelty to Animals
UNESCO	United Nations Educational, Scientific and Cultural Organization
WWF	World Wildlife Fund/World Wide Fund for Nature

Introduction

The plan of the book

The central concern of this book is an examination of support for Friends of the Earth (FoE) and Amnesty International in Britain.[1] It compares survey data gathered on these groups in 1993 with surveys of political participation in other UK political organizations (e.g. political parties), and with data from North America on interest group membership. More generally this book looks at those 'mail order' (Mundo, 1992: 18)[2] groups that have grown through heavy advertising expenditure, sophisticated marketing techniques and direct mail solicitations.

Why do people choose to join organizations such as FoE and Amnesty (in preference to political parties)? Who joins? What motivating factors are relevant? What is the significance of this decision? Why do some individuals choose to remain in membership while sizeable proportions leave? Is the support for campaigning causes an irresistible and growing wave or has it entered a period of stagnation or decline? What is the significance for democracy of this sort of participation?

A second aim of the book (begun in chapter 1) is to introduce, review, and assess the exploding literature on the environmental movement. The book deliberately tries to marshall very different approaches. This may have costs in terms of thematic consistency, but the hope is that there is a value in reading across different approaches. As we made progress in this area we became increasingly aware of the consequences of taking too broad a brush in dealing with these phenomena. The concept of an environmental 'movement' needs to be proved rather than assumed. We join others such as Kriz (1990) and Bosso (1994) who see the variations in the 'shades of green' as more significant than the similarities. The idea of a coherent movement may be illusory.

Chapter 2 essentially offers a critique of the perspective which presents organizations such as FoE and Amnesty within the *new social movement* (NSM) approach. Placing groups such as FoE and Amnesty within such a paradigm has important consequences for our perception of their role in the political system – it leads towards the conclusion that they enhance participatory democracy in the United Kingdom. Our perspective is that such *assertions* need to be back up by stronger evidence. The chapter introduces the social movement literature to discover what 'added value' it gives to a traditional political science approach. Our stark answer is 'little'. (Because its conclusions are negative this chapter is optional for the reader interested in the general issues of the book as opposed to the specialist academic debate.)

Chapter 3[3] challenges the dominant (political science) paradigm explaining membership and mobilization of interest groups (and other organizations) in terms of the provision of *selective* incentives to encourage their members to join, or indeed, to stay in membership. It highlights the existence of *public interest groups* as a theoretical puzzle for an Olsonian-type (economically driven) rational choice perspective (Olson, 1965) as public interest groups are attempting to secure *collective* rather than selective rewards. The chapter systematically reviews different theoretical approaches that challenge this appearance of paradox, arguing that rational choice analysis must subsume non-material incentives. It also highlights that discussions predicated on conceptions of rationality other than that used by Olson do not necessarily imply that participation is a problem to be explained.

Chapters 4 and 5 put the empirical 'flesh on the bones' of the theoretical discussion contained in chapter 3 with data from both our own surveys and those carried out by others. Chapter 4 provides socio-demographic data on the memberships of FoE and Amnesty. It juxtaposes our data with various US published projects; and with data on non-public interest group that we surveyed in the United Kingdom whose members join 'solely' for the selective incentives the group offers, i.e. the National Farmers' Union (NFU) Countryside. Chapter 4 finds a remarkable similarity in terms of the socio-demographic profiles of the membership of the organizations surveyed (e.g. class, education, income, and occupation). The major difference between the memberships of the (primarily) 'non-selective incentive' organizations and (primarily) 'selective incentive' organizations is that the former tend to be female, centre-left voters, while the latter tend to be male, centre-right voters. Chapter 5 examines the reasons that might account for both the *similar* (socio-demographic) and *dissimilar* (ideological) pattern of membership reported in chapter 4 between FoE/Amnesty and NFU Countryside.

Following the discussion of theoretical approaches in chapter 3 and of empirical findings in chapters 4 and 5, chapter 6[4] argues that strict rational choice analyses of interest group membership have failed to give sufficient emphasis to the role of groups in manipulating the joining decision. It moves the focus from *demand-side* to the *supply-side*. It maintains that active marketing by the group can exploit sympathy for causes and encourage the *predisposed*[5] potential member to join: a decision which might not have been faced without group prompting. Groups encourage awareness about an issue; seek to convince potential members that the group objective is attainable; and can manipulate the effective (and/or perceived) cost of an annual contribution downwards by seeking funds from other sources (e.g. patronage) to subsidize membership, and by offering selective incentives (e.g. free gifts or services exclusively to members).

Finally, chapter 7 examines the implications of the evidence in the six previous chapters. It examines the issues of group *density* and *diversity* in trying to establish whether the scale of recorded environmental activity in the political system has altered the balance of group strengths. There has recently been attention to renewed participatory democracy (e.g. Goldsteen and Schorr, 1991). Chapter 7 also examines 'modes of participation' and tries to gauge the importance of this sort of group-based extension to the democratic repertoire. It analyses the issues of the decline of party (and replacement by group) thesis, and the way in which the need to maintain support for financial reasons constrains the policy-making options available to groups.

The surveys

The membership profiles reported in the book are derived from a series of postal surveys conducted in 1993 (funded by an ESRC grant – R 000 23 3025). Amnesty and FoE were selected for three main reasons: first, they are public interest groups[6] attracting large scale memberships (100,000 plus) which challenge the Olson (1965, 1971 edn) assumption about the importance of material selective incentives in accounting for membership; secondly, arguably they can be perceived as relevant *policy participants* (Jordan, Maloney and McLaughlin, 1992); and thirdly, involvement in such groups is itself a form of democratic participation. FoE and Amnesty are major examples of a type of political activity commonly held to have increased.

We surveyed 500 Amnesty International (British Section) members, and 1,500 supporters and ex-supporters of FoE Limited based in London.[7] The sample names and addresses were provided by the organizations. Our FoE sample was divided into three main sections: 500 veteran members who had been in membership for over one year; 500 new members who were in the first year of their membership; and 500 lapsed members who had failed to renew their membership subscriptions. We then created a fourth category of 'current members' by combining and weighting the veteran and new samples. The response proportion was 55:45 veteran:new. However, the actual ratio of veteran to new members in FoE is 82:18 and consequently the samples were weighted in a way that permitted analysis of all 'active' FoE members. This category will be referred to as the 'weighted FoE sample'. The FoE membership was divided into veteran and new sub-categories because we were interested in the issues of organizational membership *and* retention. The factors encouraging long-term membership are as important as those that encouraged individuals to allow their membership to lapse. In addition, with samples of veteran, new, and lapsed members factors causing turnover could be examined.

The surveys were carried out between March and July 1993. The response rates were: Amnesty = 75.2 per cent (n = 376); FoE lapsed members = 52.4 per cent (n = 262); FoE new members = 62.8 per cent (n=314); FoE veteran members = 74.0 per cent (n = 370) and FoE weighted sample = 68.4 per cent (n = 684).[8] However, the number of questionnaires analysed was slightly less than indicated above because several were returned some six months after reminders. The analysed response rates are: Amnesty = 72.4 per cent (n = 362); FoE lapsed members = 47.6 per cent (n = 238); FoE new members = 62.6 per cent (n = 313); FoE veteran members = 73.6 per cent (n = 368); and weighted weighted sample = 68.1 per cent (n = 681).

The book also draws on several surveys conducted by the groups themselves. For example, FoE carried out a survey of its members, monthly standing order (MSO) contributors, donors and lapsed supporters in March 1992.[9] Amnesty surveyed 4,000 of their 'lapsers'. This data provides a useful comparison/check on the data collected in our 1993 surveys.

Both Amnesty and FoE are high-profile organizations which need little description (but some basic information is appended to chapter 1). Internationally Amnesty claims over a million members world-wide. Amnesty International (British Section) has around 100,000 members and FoE (UK) has around 112,000 supporters.

We have also included in chapters 4 and 5 data from our survey of 400 members of NFU Countryside (*analysed* response rate = 54.8 per cent [n = 219]). As the name suggests this has an environmental interest in a broad sense but it is closer to the Olson 'ideal type' in terms of an organization attracting a membership through the provision of material selective incentives (most significantly, competitively priced insurance). NFU Countryside attracted 33,500 members by August 1996. The full membership of the NFU (England and Wales) stands around 100,000.

Among those who worked on this project were Andrew McLaughlin, Gail Nicoll, Anita Morrison and, above all, Hettie Ras. Various colleagues in the broad academic community have commented on parts of this work. These include Stanley Cohen, Keith Dowding, Wyn Grant, Alastair McCulloch, David Marsh, Wolfgang Rüdig and, most importantly, Lynn Bennie, who commented on numerous drafts of most of the chapters in this book. We are grateful for the assistance of staff at various interest groups – most particularly those at FoE and Amnesty who helped produce the samples on which the central chapters of this book are based. These include: Sybil Clarke, John Llewellyn, Joanne Smith and Peter Wright (FoE); David Christopher and John Baguley (Amnesty); and Anne Whitehouse (RSPB). None of the above bear any responsibility for the conclusions.

Notes

1 Hereafter Friends of the Earth will be referred to as FoE, and Amnesty International British Section as Amnesty.

2 Hayes (1986: 136) refers to the same idea with the term 'Check-book members'.

3 This chapter is based on a paper published in *Political Studies* (44(4) 1969).

4 This chapter is based on a paper to be published in the *British Journal of Political Science*.

5 Groups can also create predisposition.

6 The essential quality of the public interest group is that it seeks goals that are not immediately of material benefit to the member. In practice such a definition neglects the important point that the group membership often get a bigger reward from success in obtaining the benefit sought because they value the outcome more highly than the general public. Thus if an organization changes legislation in a pro-life direction much of the broad public would not value the result as much as activists, and indeed many would see the move as detrimental to their own goals.

7 FoE (Scotland) is a separate organization based in Edinburgh but some of the members (technically supporters) of the London-based organization reside in Scotland. Following FoE's own use we will sometimes term supporters as members.

8 The response rate may be attributable to the relatively short questionnaire we sent out – fourteen pages and twenty-six questions, excluding sub-questions – and the fact that members, by definition, have some interest in the topic.

9 FoE sent out 7,500 questionnaires to members and donors respectively and received a response rate of 32 per cent for members and 31 per cent for donors; 5,000 questionnaires to MSOs with a 40 per cent response rate; and 15,000 questionnaires to lapsers with a 13 per cent response rate.

1

The inexorable growth of environmentalism?

(We) are not merely friends of the Earth – we are its guardians and trustees for generations to come. The core of Tory philosophy and the case for protecting the environment are the same. No generation has a freehold on this Earth. All we have is a life tenancy – with a full repairing lease. And this Government intends to meet the terms of that lease in full.

(Margaret Thatcher; cited in McCormick, 1991: 60)

It is widely assumed that the rise of the environmental movement has radically and irreversibly transformed the political agenda in the last 25 years or so[1] The new orthodoxy is that groups are numerous and increasingly well supported. For example, a Department of the Environment (DoE) report in 1992 claimed that membership of the main environmental organizations in the United Kingdom can be used as an indicator of committed concern about the environment. Eighteen per cent of the English respondents to the DoE survey conducted in 1991 claimed membership of an environmental body (13 per cent in Scotland). The organizations mentioned most were the National Trust,[2] Royal Society for the Protection of Birds (RSPB), World Wildlife Fund (WWF), Greenpeace, and FoE.

The rise in political importance of environmental issues is such that in 1989 a MORI poll[3] found that: 'Only law and order and the health service were seen as more serious issues than "pollution and the environment". When the public was asked to choose from a list of 24 issues (arranged in alphabetical order), law and order rated 52 per cent on the scale of concern and the health service 44 per cent. A category "pollution and the environment" was rated at 39 per cent' (Jacobs and Worcester, 1990: 37). *Eurobarometer* (a regular survey on behalf of the European Union) reported in issue 40 (December, 1993) that across the 9 countries in the EC in 1974 the environment was rated the top problem by 6 per cent and this had risen to 12 per cent by 1993. In the United Kingdom the comparable figure rose from 15 to 29 per cent.

This book looks centrally at the membership of FoE in Britain and investigates the motivations underpinning environmental mobilization. Data on members of Amnesty is also introduced in an examination of how far FoE resembles, or differs from, other large-scale campaigning public interest groups. It also draws liberally on information on other political systems, easily the best documented of which is the United States. The United States data on the apparent strength of pro-environmental values among the public *appears* even more clear cut than that in the United Kingdom. An Environmental Opinion Study of 1991 found that nine out of ten Americans were willing to identify themselves, at least weakly, as environmentalists. Bosso (1991: 156) cites a Gallup poll in 1990 which found that 76 per cent of Americans considered themselves as environmentalists. Even more surprising was that in a *New York Times*/CBS poll 74 per cent of respondents agreed with the statement (which Bosso said was regarded as zealous even by many environmentalists) that: 'Protecting the environment is so important that requirements and standards cannot be too high, and continuing environmental improvements must be made regardless of cost.'[4]

On the other hand Dunlap (1992: 99) has shown that the United States evidence is ambiguous. He notes that several environmental ballot measures, such as California's 'big green' initiative in 1990 failed (1992: 107). A nation-wide exit poll during the 1992 presidential election found the environment was ranked *last* of nine issues that had influenced voting. It was advanced by 6 per cent of the electorate compared with 43 per cent who suggested 'jobs and the economy'. However, as Scott and Willits (1994: 240, quoted in Johnson, 1995) observe: 'although people express a relatively high level of concern about the environment, they engage in few environmentally oriented behaviours'. The contradictions of opinion polls make it unwise to simply *assume* that either those expressing 'green' pro-environmental views are in any real sense part of any Green movement or that different sorts of worries aggregate into a coherent whole. As Nas (1995: 276) points out: 'The term "green" is used to describe a heterogeneous group of people and organizations, varying from those protecting wildlife to those who believe that society should be fundamentally changed.' Following Dobson

(1990) Nas distinguishes between *ecologists* and *environmentalists*: 'environmentalists try to solve environmental problems within the existing social, political and economic system, while ecologists are convinced that a fundamental change in the dominant way of thinking, living, and producing is needed' (Nas, 1995: 277).

While opinion poll data is not necessarily an accurate measure of societal attitudes, and the levels reported fluctuate, nevertheless the underlying trend of support for environmentalism has undoubtedly grown compared with the 1950s. The environment is a mainstream political issue like the economy or unemployment. The high profile of environmental issues both reflects and causes high levels of public support for pro-environmental organizations. Public support for environmental groups legitimates political claims via the political 'clout' derived from large numbers, and it also increases the financial muscle of these organizations.

As a sign that environmentalism is now a major force in British, European and North American politics it is very difficult to identify any policy development which is not given an environmentally friendly veneer. The extension of value added tax to domestic fuel in Britain was defended by the government for its conservationist implications. In June 1995, Shell (though the switch in policy was opposed by the British government) was forced to abandon deep-sea disposal of an oil installation – the Brent Spar – after international pressure by Greenpeace (*Sunday Times*, 25 June). As Witherspoon (1994: 107) points out, the growth of environmentalism to this extent has led commentators such as Mary Douglas to dub environmentalism a new *secular religion*. As Rüdig *et al.* (1993: 21) point out: 'In 1988 and 1989, all established parties were scrambling to obtain a green profile.' In a speech to the Royal Society (27 September 1988) Margaret Thatcher discussed the problems of population growth, global warming, ozone depletion, and acid rain, arguing that her administration supported the concept of 'sustainable development'. She said: 'stable prosperity can be achieved throughout the world provided the environment is nurtured and safeguarded. Protecting the balance of nature is therefore one of the great challenges of the late twentieth century' (cited in McCormick, 1991: 60).

Some commentators argued that the then Prime Minister's speeches marked a watershed which moved environmental issues from the margins to the political centre. Jonathon Porritt – who was

FoE Director when Margaret Thatcher made the speech – said that: 'This was not an easy speech for Mrs Thatcher to make; only a couple years ago she was referring to environmentalists as "the enemy within" and now she will be hard put to deny suggestions of a major *volte-face*. What's more, given the Government's record to date, people are unlikely to settle for fine words' (*The Times*, 29 September 1988; cited in Flynn and Lowe, 1992: 26). However, by the end of Margaret Thatcher's tenure in office the 'Damascusian conversion' appeared to have had more to do with halting the electoral rise of the Green Party, than a 'true' green conversion. As the *New Scientist* of 20 May 1989 put it: 'politicians have a habit of jumping onto moving trains, then claiming to be driving them' (cited in Rawcliffe, 1992: 8).

Reinforcing the data on the growth of environmental groups as an apparent indicator of environmental expansion of interest was the emergence of 'green' voting. In Britain in the 1989 European elections the Green Party achieved a 'flash vote' of 14.9 per cent. Franklin and Rüdig's (1993: 129) analysis of that election concluded that: 'green voting is more common among the young, well-educated, and middle class, and green voters also tended to be left-wing, post-materialist and concerned about the environment and arms limitation'. Franklin and Rüdig also claimed that this summary is applicable to green voting patterns in Germany and the Netherlands. Ultimately, the data on voting erodes rather than sustains the view of a relentless rise of green awareness. Green voting, in the United Kingdom at least, has been characterized by an exceptionally high degree of volatility: there is little evidence of sustained change.

Nevertheless, Dalton (1994: xiii) has argued against the idea that green politics is a passing novelty claiming that the failure to connect more fully to environmental awareness may have contributed to the downfall of George Bush and Margaret Thatcher (Dalton, 1994: 51). Viewed from 1996 this verdict appears to exaggerate the robustness of the environmental issue. More recent evidence discussed below shows at best a 'plateau' of environmental concern – if not a decline. As Rawcliffe (1992: 8) put it: 'After the *Greenrush*' of the late 1980s in Britain 'much of the fire has ... gone out of the environmental debate'.

Environmental phases: the rise of environmentalism[5]

Dalton (1994: 26) categorized the history of environmentalism in two broad phases.[6] The *first* at the turn of the century dealt with wildlife protection and preservation of the physical fabric in terms of landscape and buildings. Dalton sees the modern nature conservation movement beginning to develop in Western Europe during the last half of the nineteenth century. The Society for the Protection of Birds was established as early as 1889 (becoming the RSPB in 1904), with other bird protection societies appearing in other continental countries about the same time (e.g. Dutch and German bodies were established in 1899). The National Trust in Britain was created in 1850 with comparable preservationist groups developing in other European countries (e.g. Association for the Preservation of National Monuments, and Heemschut, in Holland (Dalton, 1994: 31)).

Dalton presents the mobilization wave of 1880 to 1910 as reaching a plateau – with the inter-war years representing a continuation of that plateau. He characterizes the modern conservation movement as a post-war phenomenon – with the Town and Country Planning Act in 1947 in Britain signalling a new impetus towards using the power of government to preserve the countryside. This period also saw international developments such as the creation of the International Union for the Conservation of Nature appearing under the auspices of the United Nations Educational, Scientific and Cultural Organization (UNESCO). (In the United States a phase that ended about 1960 was identified by Costain and Lester (1995: 25) as the Conservation-Preservation Movement (1920–60). The attention to the issue appears to have been in large part about protecting the leisure pursuits of the upper middle class.)

Dalton identifies the late 1960s as starting a distinct second (new) wave of European environmentalism. This he partly ascribes to the problems being experienced by advanced industrial democracies – nuclear power, resource shortages, toxic waste, acid rain, etc. It was also a response to the educational efforts of conservationists – most notably Rachel Carson's *Silent Spring* (1965) and Barry Commoner's *Science and Survival* (1963). Kimber and Richardson (1974) noted that between 1957 and 1973 the number of local groups registered with the Civic Trust in the United Kingdom rose from 200 to 1,000. It has been claimed that the groups started in this broad period were of a different character from earlier cre-

ations. According to Nas (1995: 275): 'they changed their orientation from conservation to a critique of prevailing methods of production and patterns of consumption'. Environmental groups of the
1960s and 1970s portrayed a gloomier picture of the 'inevitable'
environmental catastrophe which could only be avoided 'by fundamental and radical changes in the values and institutions of industrial societies' (Cotgrove, 1982: 3). [7]

The second stage of development has (arguably) led to an
increased scale of activity in two distinctive ways. First, there was
the expansion in the number of environmental organizations in the
period 1966–75 (Lowe and Goyder, 1983). For example, the 1994
Directory for the Environment contained over 1,600 organizations
(Frisch, 1994). Particular components of the environmental debate
threw up large numbers of groups. For example, the transport issue
has seen the emergence of groups such as CHARM (Cyclists Have
a Right to Move), Reclaim the Streets, Revolutionary Pedestrians'
Front, as well as longer established groups with an interest such as
the Pedestrians' Association and the National Society for Clean Air
(*Sunday Times*, 30 July 1995).

Secondly, there has been an expansion in the *membership* of environmental groups, both in the United Kingdom and United States.
For example, in 1981 FoE (United Kingdom) had 18,000 members;
by 1991 it had 114,000 (+533 per cent). In the same period, the
RSPB's membership grew from 441,000 to 852,000 (+93 per cent),
while Greenpeace (United Kingdom) increased its membership
from 30,000 to 408,000 (+1,260 per cent).[8] In the United States, in
roughly the same period (1980–90), the Sierra Club grew from
181,000 to 630,000 (+248 per cent), the National Audubon Society increased from 400,000 to 575,000 (+44 per cent), and at the
zenith of the membership explosion (1985–90) Greenpeace (USA)
grew from 800,000 to 2.35 million (+194 per cent) (see below,
Tables 1.1 and 1.2).[9]

In 1992 the total membership of United Kingdom environmental
groups was estimated at 5 million (or 9 per cent of the population)
(Grove-White, 1992: 128; cited in Rüdig, 1995). While in the
United States Johnson (1995: 1) estimated that the total number of
members of environmental organizations was around 15 million.
Thus, the usual claim is that not only has the number of groups
mushroomed but so have the numbers of those who have, to this
extent, become involved in politics.

Table 1.1 *Membership trends among selected UK environmental and campaigning organizations (000s)*

	1971	1981	1984	1986	1987	1988	1989	1990	1991	1992	1993	1994
Civic Trust[a]	214	–	247	241	240	244	293	293	222	222	222	222
Conservation Trust	6	5	5	–	–	3	3	3	1.6	–	–	–
Council for the Protection of Rural England	21	29	–	–	32	32	40	44	45	46	45	46
Friends of the Earth[b]	1	18	27	28	55	65	140	110	114	116	120	112
Greenpeace	–	30	–	–	–	–	320	380	408	411	410	300
National Trust[b]	278	1,046	1,328	1,417	1,545	1,634	1,865	2,032	2,152	2,186	2,189	2,219
Ramblers Association	22	37	44	53	57	65	75	81	87	90	94	100
Royal Society for Nature Conservation	64	143	161	179	184	204	210	250	204	250	248	–
Royal Society for the Protection of Birds	98	441	466	506	561	540	771	844	852	850	850	870
World Wildlife Fund for Nature (UK)[c]	12	60	76	106	124	147	200	247	227	209	207	187

Source: Social Trends, 15, 16, 18, 19, 20, 21, 22, 23, 24, 25 and 26.
Notes:
[a] Members of local amenity societies registered with the Civic Trust.
[b] England and Wales and Northern Ireland only: separate organization for Scotland.
[c] These figures exclude an additional one million 'other' supporters and donors who are non-members.

The increased prominence of the environment as a political force has been accompanied by structural differentiation – not all groups have grown in membership terms. In the United Kingdom organizations such as FoE and Greenpeace have, to a certain extent, flourished at the expense of smaller and/or more local environmental groups (e.g. the Conservation Trust) (see Table 1.1). Somewhat ironically, in the United States it has been FoE which has been 'squeezed' (out) by more mainstream competitor organizations.

Table 1.2 *Membership trends among selected US national groups, 1970–1992 (000s)*

	1970	1980	1985	1990	1992
Sierra Club (1892)	151	182	364	630	650
National Audubon Society (1905)	105	400	550	575	600
Izaak Walton League (1922)	54	52	47	50	53
Wilderness Society (1935)	54	45	147	350	313
National Wildlife Federation (1936)[a]	540	818	900	997	975
Defenders of Wildlife (1947)	13	50	65	75	80
Nature Conservancy (1951)	22	n.a.	400	600	690
World Wildlife Fund (1961)	n.a.	n.a.	130	400	940
Environmental DefenseFund (1967)	11	46	50	150	150
Friends of the Earth (1969)[b]	6	n.a.	30	9	50
Environmental Action (1970)	10	20	15	20	16
Greenpeace USA (1972)	n.a.	n.a.	800	2.35m	1.8m

Source: Bosso, 1995: 104; Mundo, 1992: 176.
Notes:
[a] Full members only. The federation in 1992 also had affiliated memberships (e.g. schoolchildren) of 5.3 million.
[b] FoE merged in 1990 with 30,000-member Oceanic Society and the non-member Environmental Policy Institute.

Comparison of the scale of FoE in Tables 1.1 and 1.2 shows that its place in the British interest group system is more prominent than in the United States. Bosso (1994: 35) says that in North America FoE is one of the smaller, narrowly focused, more brethren-type organizations compared with large, multi-issue membership groups such as the National Wildlife Federation. The success of groups seems at least in part to be determined by the competition they meet. In Britain there were not perhaps as strong general purpose environmental organizations with which to compete, thus FoE was able to take on that role. Moreover, the 'purist' niche of the market was perhaps occupied by Greenpeace (United Kingdom): thus in the two countries rather different styles of organization evolved.[10]

The growth in membership of environmental and campaigning groups has seen an increase in these organizations' financial resources. MORI polls have shown that in the United Kingdom between 1988 and 1994 the number of people donating money to environmental causes has grown substantially, peaking in mid-1991, but the current levels (around 55 per cent of respondents claiming to have contributed to an environmental cause) are substantially higher than the mid-1988 figure of around 27 per cent.[11] Rawcliffe (1992: 3) estimated that: 'In 1990, the United Kingdom's 15 largest national environmental groups, including the National Trust, had a combined annual budget of £163m.' While in the United States, Letto (1992: 28; cited in Bosso, 1995: 102) estimated that in 1990 'the environmental community as a whole took in some $2.9 billion – almost double 1987 revenues'. [12]

In the United Kingdom in 1994–95 FoE's income was approximately £5.3 million and it employed 96 staff (*Independent on Sunday*, 5 May 1996). While in 1990 Greenpeace (United Kingdom) had 54 full-time staff, World Wildlife Fund for Nature (United Kingdom) 160, RSPB 428, and the National Trust had 1,908 (Cowell, 1990). As Table 1.3 and the data above illustrate many environmental organizations are on a par with small to medium-scale businesses. [13]

Table 1.3 *Budgetary trends among selected US national groups, 1970–1992 ($millions)*

	1970	1980	1985	1990	1992
Sierra Club	3.0	9.5	22.0	40.0	41.0
National Wildlife Federation	13.0	34.5	46.0	89.5	90.0
Defenders of Wildlife	n.a.	n.a.	3.0	4.3	4.4
Nature Conservancy	n.a.	n.a.	156.0	200.0	216.0
Environmental Defense Fund	n.a.	2.0	3.5	16.0	20.8
Friends of the Earth	0.36	1.0	1.0	3.0	3.3
Environmental Action	n.a.	0.55	0.6	1.1	1.3
Greenpeace USA	n.a.	n.a.	24.0	35.0	50.0

Source: Bosso, 1995: 105.

Some environmental groups have elevated fund-raising and organizational marketing to a major (in some cases the major) group activity (see Figures 1.2 and 1.7). In 1992 it was reported that FoE had approximately 30 staff on campaign appeals and membership drives, and the same source argued that much of the Council for the Protection of Rural England's (CPRE's) membership growth in the mid-1980s can be attributed to a recruitment drive organized by an advertising agency, Wilmot and Partners (Rawcliffe, 1992: 3). As we (in chapter 6), Bosso (1995), and Godwin (1988) argue, the growth in size of environmental organizations cannot be solely attributed to the increased saliency of environmental issues. The sophisticated marketing efforts of these large-scale organizations has had a significant effect on group size: creating an *activated* constituency.

Growth, plateau or decline?

The figures on group membership and resources noted above are an attempt to add a sense of scale to the popular and broad claim that 'there is a lot of this about'. However, an accurate and full picture would not simply record a steady increase in the scale and *organizational* success of environmental and other public interest groups: it would highlight the significant membership turnover problem faced by these groups (see chapter 6). Consequently, it could be argued that public interest groups with large memberships are endemically under-funded and hence weak in the group contest – are the sorts of incentives that attract membership of trade associations and other material interest groups more likely to guarantee rejoining? The incessant pursuit of new members is a life-preserving activity for environmental and campaigning organizations. Turnover rates are very high, in many cases averaging between 30 and 40 per cent in both the United Kingdom and the United States (Letto, 1992; Lowe and Goyder, 1983; Cohen, 1995).[14]

The instability in group aggregate membership has in the past been masked by the success in recruiting that has not merely replaced those leaving but also seen an overall increase in support – although membership growth appears to be plateauing out for some groups and in decline for others across the public interest sector in the 1990s (see Tables 1.1 and 1.2). For example, membership data from FoE in 1995 records that membership appears to have 'peaked' in 1991. FoE uses two main categories of 'membership': 'members, covenantors/payroll givers' and 'donors' (see Table 1.4). It is the regular contributor that appears to be a particular problem for the organization. It has been claimed that subscription income for Greenpeace fell by half in the 1990s – partly due to world recession. From a peak of 4.8 million supporters internationally, it lost 1.7 million by early 1995 (*Scotland on Sunday*, 8 September 1996).

Table 1.4 *FoE membership*[15]

	1989	1990	1991	1992	1993	1994
Members, covenantors and payroll givers	75,000	115,000	120,000	92,921	83,111	79,860
Donors	50,000	50,000	55,000	61,790	51,353	42,621

In the past decades there have been conflicting perspectives on the growth of environmental support. On the one hand it is seen as ever-expanding and rippling through the population as the environmental explosion takes effect. Baumgartner and Walker (1988: 909) report for the United States that membership of the five most prominent environmental groups rose from 439,000 in 1966 to 1,217,600 in 1975 – 'and there was no sign that the trend was slackening'. There is a second view that sees a finite public attention span to any topic. In 1972 Downs set out his model of the 'issue-attention cycle' and he then predicted that the environment would, like other issues, have high-profile prominence and then be overtaken by fresher topics. There is a third view that sees environmentalism as a 'sunshine issue' that emerges only when economic improvement permits this 'luxury' (Dalton, 1994: 51). Dalton (1994: 72) concluded that the consistency of public support through the recession of the 1980s proves that environmental concerns are not ephemeral. Although the membership data in Tables 1.1, 1.2, and 1.4 suggests that there is at least as much evidence of stagnation or decline as there is of inexorable growth.

Organizational transformation? The rise of protest businesses

Arguably a more significant type of change than the increase in the number and membership of groups concerns the *style* and *structure* of these organizations. Our core argument is that as environmental organizations have greatly increased the scale of their support, they have also transformed the implications, requirement, and *nature of that support*.

These groups are often seen as an integral part of the (environmental) New Social Movement (NSM) and certain implications are commonly held to follow from such a categorization/perception. These we question. While Dalton (1994: 9) concedes that the framework for studying NSMs is often imprecise, he argues that the general characteristics appear to include a decentralized, non-hierarchically structured organization that 'reflects the participatory tendencies of their members'.

Our argument is that this sort of image is now (if ever) applicable to only a small part of the so-called environmental movement. In the United States, Salazar (1995: 5) has claimed that as environmental concerns have been institutionalized in laws and public agencies, environmental organizations have been transformed into professional, bureaucratic, mainstream interest groups. She says the major national groups are now led by managers, staffed by professional economists, lawyers and biologists, and supported by sophisticated public relations and fund-raising departments. In Britain, Rawcliffe (1992: 3) discusses the rapid maturing of the environmental groups as political actors in the 1980s and early 1990s. He, like Salazar, notes their transformation into corporate organizations: 'with large membership and sponsorship income, management structures and networks, scientific research capabilities and sophisticated public relations machines ... these developments have been hallmarked by the increasing professionalism of the United Kingdom's environmental lobby'. This is very different from expectations about environmental movements that stress their low level of central organization.

Some parts of the environmental terrain at least claim to be leaderless. A poster (1996) in Aberdeen for the cycle group 'Critical Mass' spells out: 'These (mass) rides operate as a *xerocracy* – rule by Xerox machine and no (zero) leaders.' A study in 1994 of the changing culture within Greenpeace in Britain in a Channel Four television documentary called the early Greenpeace an 'anti-institution'. But as suggested by the programme's title – *Greenpeace: Where Have the Warriors Gone?* – the programme's argument was that the group had been bureaucratized and routinized over time. Greenpeace veterans were quoted seeking 'more spontaneity, more risk, less bureaucracy, quicker access to decision takers'. In 1995 (*The Observer*, 10 December) the former Executive Director of Greenpeace United Kingdom, Allan Thornton, complained that Greenpeace International's growth in finance was not accompanied by a parallel increase in campaigning effectiveness. He argued that the organization spent 'more and more' of its time on long-term strategy meetings, with the most effective campaigners being forced to become managers: '... much of the talent attracted into the national offices was in marketing and fundraising. *Big offices, big budgets and big boats did not necessarily make for better campaigns'* (emphasis added). In 1992 a group of Greenpeace's own campaigners hijacked the *Sirius* in Amsterdam and took it out to sea as a protest against the organization's drift from its early goals.

FoE has been subject to a similar critique from some of its founding members. For example, an article in the *Independent on Sunday* (5 May 1996) quotes Peter Wilkinson (who ran FoE's first campaign on non-returnable bottles):

dollar for dollar, it is achieving less change now than in the early days ... In those days we used to say that if we had a quarter of a million supporters, an income of £1 million a year and a high profile we could revolutionize the country. All of the benchmarks have been exceeded and great credibility built up and yet the new world is as far away as it was in 1971. I do not think Friends of the Earth has achieved as much as it should have done.

Whatever the mythology of 'anti-organizations', Amnesty, FoE and Greenpeace are similar in that like many other large-scale public interest groups they operate along corporate lines. Amnesty and FoE adopt a business strategy to ensure efficient use of resources, to maximize effectiveness and, probably above all, to ensure organizational survival in a highly competitive market (see Table 1.5).

The business orientation of many 'new politics'[16] groups is not, however, without its defenders. Rawcliffe quotes Grant Thompson of the Conservation Foundation in the United States, who was told by the auditor of an organization he was involved with that:

You should never forget ... that you're in the good works business. If you don't pay attention to the business, you won't be doing the good works very much longer.

(Rawcliffe, 1992: 5)

Rawcliffe (1992: 5) argues that the liquidation of the Earthlife Foundation in 1988 and the demise of Ark clearly illustrate the *Thompson principle*. Stanfield (1985: 1350) noted the leadership changes in a number of US groups, arguing that these changes had occurred largely because the increased size of the groups changed the demands and requirements of those in the leadership posts. He quoted the Director of the National Audubon Society (which had grown by 70 per cent to 550,000 in the first half of the decade): 'the

Table 1.5 *Selected income and expenditure data for the charitable parts of selected UK environmental and campaigning organizations (£millions)*

	Total voluntary Income	Covenants	Legacies	Other donations	Fund-Raising income	Trading	Total income	Fund-raising expenditure	Administration expenditure	Total expenditure
Amnesty International	1.057	0.527	0.167	0.363	–	–	1.106	–	0.017	1.192
Friends of the Earth	1.440	0.632	0.123	0.685	–	–	1.936	0.331	0.154	1.955
Greenpeace	1.585	0.060	0.452	0.988	0.085	–	1.627	0.052	0.084	1.525
National Trust	65.207	27.705	22.259	15.243	–	7.029	132.355	13.607	5.536	118.725
Ramblers Association	1.285	0.272	0.070	0.760	0.183	0.016	1.537	0.250	0.231	1.490
Royal Society for the Protection of Birds	21.883	1.240	7.234	12.188	1.221	1.852	27.800	3.827	0.976	25.129
World Wildlife Fund for Nature (UK	15.029	–	3.708	9.792	1.529	1.638	18.981	4.564	1.075	19.869

Source: Charities Aid Foundation (1994), *Charity Trends 1993*, 16th edn.
Note:
The figures relate to the charitable wings of some organizations.

people at the top (of the groups) found they were running a big business and may not have been comfortable ... Some of them enjoyed substantive advocacy rather than the nitty-gritty of managing a business' (Stanfield, 1985: 1350). By 1995 British groups such as the National Trust were advertising a post of Managing Director of National Trust (Enterprises) Ltd. A salary of £60,000 plus benefits was on offer. The post holder was expected to 'maximize income, through the development of new ideas, innovative products and programmes' for the Trust's 200 shops, plus catering outlets, holiday cottages, licensing and affinity agreements and a mail order business. By 1995 the *Sunday Telegraph* (22 October) was claiming that the typical salary of a top fund-raising charity was £70,000. It said that the combined salaries of the National Trust's nine management board members was over £500,000: 'Most charity directors now have chauffeured or company cars.'

In fact, it is the desire for aggregate membership growth in spite of the turnover problem noted above which has required the recruitment of top business executives. For example, the Defenders of Wildlife in the United States largely lost out on the membership surges of the 1980s and experienced a higher than average turnover rate. In response to this it appointed Rodger Schlickeisen as its President in 1991. He had an economic and budgeting background and had previously worked at the Office of Management and Budget under Carter, as chief executive for Senator Max Baucus, and as chief executive officer at Craver, Matthews, Smith & Company – a major direct mail and telemarketing firm whose clients included Greenpeace, the Sierra Club, the Natural Resources Defense Council (NRDC), and the Defenders of Wildlife (Bosso 1995: 116). George Medley, with a background in business, became director of the World Wide Fund for Nature (United Kingdom) in the early 1980s. He set the single objective that it 'should raise the maximum funds possible from United Kingdom sources and ensure that they are used for the benefit of conservation'. *The Observer* (17 December 1995) recorded that over 12 years, the fund's income grew from less than £500,000 to more than £20 million a year. Thus it is organizational skill not environmental commitment which is sought if the group has a dependency problem on supporter cash.

The fact that organizations such as FoE have become mass-based, heavily dependent on supporter contributions (85 per cent in 1994) implies an organizational style. FoE spent 24 per cent of its 1994 income on supporter-based activities (8 per cent on fund-raising; 8 per cent on administration; 8 per cent on supporter recruitment and servicing). Bosso (1995: 111) reported that one 'conservative estimate' put Greenpeace USA's expenditure on fund-raising alone at 20 per cent of it annual budget, while Horton (1991) estimated that 'forty-eight cents out of every dollar raised through direct mail goes straight back into more fund raising' (cited in Bosso, 1991: 111). These 'successful' *protest businesses* would not spend so much time, effort, and financial resources to encourage greater mobilization, and to keep members *in* once they've crossed the threshold, were it not 'profitable' to do so. As Bosso (1995: 113) concluded: 'It is also no surprise that the most aggressively expansionist groups have been those willing to commit to sophisticated direct-mail campaigns as the backbone of their fund raising' (see Table 1.6 for the characteristics of protest business).

Table 1.6 *The characteristics of protest businesses*

(i)	Supporters rather than members are important as a source of income.
(ii)	Policy is made centrally and supporters can influence policy primarily by their potential for exit.
(iii)	Political action is normally by the professional staff rather than the individual supporter or member.
(iv)	Supporters are unknown to each other and do not interact.
(v)	Groups actively shape perceptions of problems by providing supporters with partial information.
(vi)	Supporters are interested in narrow issue areas. Particularity rather than ideological breadth is the agency of recruitment.

Bosso (1995: 101) notes that groups deliberately prefer to create a membership base as a source of finance, rather than rely on foundation or corporate money, precisely because the large numbers of small contributors 'give the group 'freedom to manoeuvre' (see Figure 1.1). The group has the numbers and resources it requires to carry the battle forward without limitations on how and/or where that money is spent. Thus, paradoxically, members free the organization from control.

Figure 1.1 *Sources of revenue for membership groups in the United States, 1993–1994*

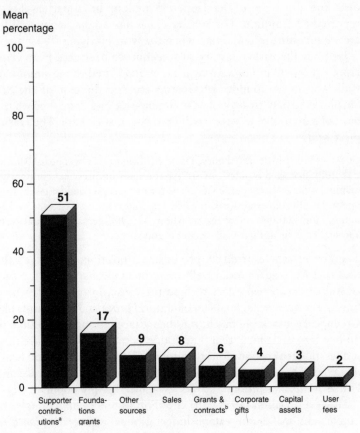

Source: Adapted from Bosso (1995: 107).
Notes:
[a] Membership dues plus individual contributions.
[b] Grants and contracts include: federal grants and contracts; state grants and contracts; and other contracts.

The *Sunday Times* (22 October 1995) drew attention to the fact that Greenpeace has transformed from a tiny band of idealistic ecologists over 20 years ago, into a corporate Goliath with an annual budget of $150 million, a staff of 1,000, and 43 offices world-wide. The article also noted that: 'The nine directors (on the international board) had a $400,000 budget for expenses, travel and private staff', and that Greenpeace had been criticized by its own activists over the purchase of its United Kingdom headquarters for a reported £2.6 million. The *Sunday Times* quoted one activist: 'Why are we investing in real estate when the world is dying?'[17]

Nor has the earlier history of Greenpeace been free from criticism. An article in *Forbes Magazine* in 1991 used the testimony of Paul Watson, co-founder of Greenpeace and director of the Sea Shepherd Society to suggest that Greenpeace had flourished on the back of a myth that it was a small David against Goliath. The *Forbes* piece said:

Under its recently departed guru, David McTaggart ... Greenpeace became a skilfully managed business, mastering the tools of direct mail and image manipulation ... Greenpeace Germany, for instance, second-largest branch operation after Greenpeace United States, had revenues last year of $36 million and 700,000 members, of whom 320,000 permit Greenpeace to automatically debit their bank accounts annually.

The *Forbes* article, from its pro-business position, challenged the idea that McTaggart was a well-intentioned idealist who left a successful career in real estate to fight the cause. Instead it presented him as a controversial financial operator. *Forbes* said that outfits like Greenpeace attack big business as being faceless and responsible to no one but, it claimed, that description better fits Greenpeace than it does modern corporations that are regulated, patrolled, and heavily taxed. Thus the image of organizations such as Greenpeace and FoE as *protest businesses* appears to be increasingly popular.

However, Dalton (1994: 19) cites FoE as one of his examples of 'unconventional or anti-establishment groups ... *generally identified as fitting the pattern of New Social Movement theory*' (emphasis added). Dalton's (1994: 40) implication is that FoE is one of the loosely ordered ecological type of phenomenon: 'displaying its populist goals in a decentralized, anti-bureaucratic organizational structure', and which pursues: 'a new model of citizen action on environmental issues ... More than just environmentalists, FoE activists were social critics ... Their tactics combined confrontation with the authorities and events with the intent of sparking public interest.' This interpretation is more appropriate for the United States version of FoE than the British.

In our view, FoE and Greenpeace are *nearer* the ideal type of *protest businesses* rather than NSM organizations or straightforward interest groups. FoE actually exists as two separate organizations – FoE Ltd and FoE Trust – and Greenpeace in Britain is more technically Greenpeace United Kingdom Ltd (from 4 November 1991), and like FoE Ltd, is a company under the 1948 Companies Act. Clause 5(c) of FoE's constitution says: 'The Directors may in their absolute discretion create one or more categories of members who shall not be entitled to vote at General Meetings and shall have such rights and privileges as the Directors may determine and a person becoming a member as aforesaid shall be called a "Friend of the Earth".' In other words, an invitation to be a 'Friend of the Earth' is not an invitation to policy-making membership of the organization, but an invitation to make a financial contribution.

In 1991/92 FoE spent £234,172 on supporter servicing, £81,322 on supporter recruitment, and £463,484 on fundraising.[19] The 1991/92 Report noted:

The year to 31 May 1991 was a difficult one for fund raising: the recession meant that people cut back on giving to voluntary organizations: the Gulf War and the number of natural catastrophic disasters meant increased competition for funds. This was counterbalanced by the increased interest in the protection of the environment during the year which saw our number of supporters [sic] grow from 200,000 to 240,000. Supporters' contributions [sic] which form our main source of income were down on the previous year by £691,845 (20 per cent) (In 1991/2 supporters contributed £3,038,338 out of a total income of £4,109,757 – £415,180 was raised by gross profit on trading) ... Fundraising expenditure also rose as a result of investment in two new areas of work (legacies and regional fundraising) which will yield returns in subsequent years.

Is this the sort of picture and language that Dalton's description of ecological groups encourages?

Environmental divisions

The label 'environmental', like most general terms, covers some very different – even conflicting – components. In 1995 the export of calves from Britain for veal production put the large number of animal welfare groups more fully into the public domain – with groups such as Compassion in World Farming (CIWF) (established 1967) using the publicity to run high-profile recruitment schemes. Such concerns are not strictly environmental but there is an overlap in the ideological and moral assumptions of some ecological and welfare groups about sustainable development and ethical issues. It is difficult to draw boundaries round groups and establish a coherent 'environmental' category. Some so-called environmental groups would seem to share the orientation of peace and sexual identity, appearing in the form of cultural groups rather than traditional conservation groups. McAdam and Rucht (1993: 73) have discussed the borrowing and links between nominally separate social movements. However, the sort of Civic Trust or National Trust membership need have little to do with *environmentalist* supporters – though in September 1995 there were reports that the National Trust was considering trying to widen its influence by being a more campaigning-type organization.

But much conservation action is particular rather than general: the local residents objecting to a relief road may be reflecting the financial implication for their own property – against a project which would lead to environmental benefits for others and an overall easing of traffic congestion. In Britain it is commonly assumed that Not In My Back Yard (NIMBY) type protest is conservative and middle class in essence. However, writing on United States developments Goldsteen and Schorr (1991: 120) maintain that these grassroots groups represent a 'new paradigm' in which communities no longer passively accept the assurances of outsiders. Bosso perceives the expansion of grassroots groups as 'environmentalism's third wave' (1992, quoted in Ingram *et al.*, 1995: 119). Kriz (1990: 1862) has argued that local protesters on local issues are often working class. She notes that in the battle against groundwater contamination from solvents (from the electronics industry): 'Local residents expected the national, mainstream environmental groups to help them pressure these big companies to stop polluting and to clean up the damage. But even national groups with high profiles in California didn't always come through' (Kriz, 1990: 1826).

Kriz (1990: 1826) quoted the director of the Silicon Valley Toxics Coalition who argued that the Sierra Club was active: 'But a lot of their membership is based in wealthy sections of the county. A lot of the traditional conservation-preservation type of environmentalists are people who live in very nice sections and have very nice jobs working for a lot of these companies that are causing all the pollution.' She also noted that the Silicon Valley coalition's support is at the other end of the socio-economic spectrum. The director was quoted as stating that: 'the people living in the flat lands ... are the ones exposed to the worst pollution ... There is class-based distinction ... There is a lot of scepticism of the mainline groups as either being the tree-huggers or the limousine liberals.'

Kriz (1990: 1826) claims that in the 1980s around 7,000 citizen groups formed to force pollution clean-ups or to prevent landfills or incinerators near their homes. These, she says, are more confrontational than their national counterparts and their strength generally lies in blue-collar, low-income areas. Consequently, many local issue activists express serious reservations about the mainstream environmental groups, which they see as too caught up in Washington power politics, too willing to compromise with industry, and not paying enough attention to local concerns. The Director of the National Wildlife Fund conceded that: 'If you're living downwind of a chemical plant, it's kind of hard to see why the national groups are so concerned about spotted owls or wilderness protection' (quoted in Kriz, 1990: 1827).

Even within the 'new environmentalists' there are significant divisions. As Cotgrove (1982: 5) points out: 'new environmentalists are agreed on the need for a fundamental social change if society is to survive. But there the consensus evaporates. On closer examination, basic differences emerge which make it possible to construct two major variations of the new environmentalism – a traditional and a radical form.'[20] Thus Greenpeace is not the National Trust for Whales; Sea Shepherd Society is not the CPRE of the oceans. Dalton's 1994 book adopts the metaphor of the *Green Rainbow* that runs from the traditionalism of the RSPB to the assertive tactics of Robin Wood (a radical offshoot of Greenpeace in Germany). But are they in fact part of the same spectrum or distinctly different?

While Dalton (1994: 49, 46) concedes that the distinction blurs, he distinguishes between a conservation orientation and an ecological orientation. He quotes Lowe and Goyder (1983: 35) who term the conservationists as: 'emphasis groups ... whose aims do not conflict in any clear-cut way with widely held social goals but which are motivated by a belief in the importance of certain values and the need for vigilance on their behalf'. Ecologists are in contrast presented as advocating a basic change in societal and political relations: 'The most important ideological feature of an ecology group is its rejection of the values of the prevailing social order. Ecologists, supposedly, are attempting to develop a new societal model – what is called a New Environmental Paradigm' (Dalton, 1994: 47). The tendency to aggregate disparate political forces into an 'environmental' category may actually mislead, but it serves both the interests of the environmentalists and their opponents to cite large (if inconsistent) numbers. 'Environmentalism' might be a collection of incompatibilities?

Dalton (1994: 46) (optimistically) suggests that wildlife preservation groups now realise that habitats and their prservation are vital and hence the interests of conservationist and ecological groups overlap. This may be the case – but at the same time the members of a conservationist group such as the RSPB may be reluctant to endorse ecological projects abroad that are too far from the immediate support of domestic species. Instead of some convergence of ecological and conservationist concerns as foreshadowed by Dalton, there may be an increasingly sharp competition over the conservationist message – particularly between groups vying to *maintain or increase their market share.*

Some opponents of the 'environmentalist' cause are now inclined to try to argue against the 'seizure' of the conservationist label by those with radical ecological beliefs. In the United States opposition to so-called environmentalists has formed under the wise use label. These voices reject bodies such as FoE (US) as being composed of ignorant and prejudiced outsiders who do not appreciate the interaction of hunting and conservation, commercial development and management, and use of federal lands. Bosso (1994: 41) says: 'Those who are part of the "wise use" cause consider themselves the *real* conservationists, lovers of the outdoors who hunt and fish, believe in scientific management of natural resources, and chafe at charges that they are despoilers of the Earth.'

A good example of the conflict between forces that would all characterize themselves as supportive of the environment concerned the development of the Warm Springs or Dry Creek Dam at Lake Sonoma in Northern California. This has been a classic battle between wise use and ecological groups. The Citizens for Community Improvement (CCI; now translated into Friends of Lake Sonoma) saw the dam as an economic necessity to protect the existing agriculture land use. They maintain that the completed dam has improved the local environment by adding recreational and conservation facilities while helping agriculture. Accordingly, the dam is perceived to be an asset. The official history of the project, *Water*

Stewardship (1983), claimed that: 'It was people at grass roots level who got the idea of Dry Creek Dam going in the 1930's ... and who led the crusade to rescue it when seriously challenged by environmental and other no-growth interests, at the election polls and in the courts of law, from about 1972 to 1980.' The dam supporters saw themselves as having a greater stake in the conservation of the area than groups who applied a few 'first principles' that saw any change as deterioration. A report in the *San Francisco Sunday Examiner* (11 April 1976) in the middle of one of the droughts the dam was meant to alleviate quoted Milt Brandt (a major figure in the history of the pro-dam group):

The people of Sonoma voted to build the dam for just such droughts. Then the shaggy-faced environmentalists who gravitate around Sonoma State College mounted a crusade to kill the dam ... It will be worth skipping a Saturday night bath this summer to know that some environmentalist who labored to kill the dam has a parched tongue dragging in the dust.

A newsletter from the CCI in 1979 referred to their opponents (the Warm Springs Dam Task Force) as 'a small group of adversaries *posing* as "friends of the environment" ... these groups were not as concerned about preserving an endangered environment as they were with preserving their rural lifestyles'. The 'environmentalists' were characterized as: 'a small, persistent segment of the rich and upper-middle-class ... [who had] taken on the banner of "environmentalism" as a front for the much less desirable goal called "looking out for number one".'

After being discussed and planned from the 1930s onwards the project was authorized in 1962 in the Flood Control Act. Construction finally started in 1967 and was under way when the 1969 National Environmental Policy Act was passed. This required that all Federal agencies analyse the potentially adverse impact of construction projects. This gave ecological-type environmental groups an opportunity to obstruct the programme. *Water Stewardship* (1983: 42) describes this intervention as follows: 'Environmental groups, aided by stop-growth advocates who feared the Dam would induce population increase attacked the Corps' Environmental Impact Study, forcing a shut down from May 1974 to May 1977.' The CCI had to fight vigorously to regain the initiative. As Bosso (1994: 41) notes the 250 or so 'wise use' groups that he identified can be mere fronts for commercial interests (e.g. timber or mining

corporations). Undoubtedly much of the support for the CCI came
from those with a business background, but there is also little doubt
that they thought that the dam, in reducing floods and drought and
improving amenities, made the Sonoma Valley a better place. Self-
interest was not, in their minds, antithetical to the public interest.

The Dry Springs story is illustrative of the point that causes seek
to *label* themselves as environmental as a deliberate attempt to
make a political point: in fact *which* proposal is more 'environmentally friendly' can often be a matter for dispute. For example, all
traffic schemes tend to be labelled by opponents as anti-environ-
mental. However, this is not necessarily so: traffic may be diverted
away from residential streets and this may lead to an easing of congestion which itself wastes fuel and causes hazardous pollution –
especially from particulates (an important public health issue). Thus
a relief road might offer environmental, safety, and public health
gains. Similarly, the decision not to sink the oil installation, the
Brent Spar, in deep ocean may ultimately be judged to be against
best environmental practice if onshore disposal leads to leaks and
other hazards, but Greenpeace was able to *assert* their opposition as
the pro-environmental stance.

The underlying argument of this section is that there can be genuine doubt about the environmental credentials of different organizations. Bosso (1995: 102) discusses the 'sharp ideological and
tactical cleavages within environmentalism'. This might be a more
realistic image than that of a coherent whole. There is such a massive potential for conflict within the so-called environmental camp
that the movement strength may be far less than the sum of the
parts. Many groups attempt to mobilize under the environmental
flag of convenience because the environmental term is 'positively
loaded' – after all who supports the destruction of the planet? Political entrepreneurs see the environmental rubric as a potentially
profitable guise.

The large battalions of environmental interest groups appear to
be organized along lines familiar to political science as formal interest groups.[21] Bosso (1991: 153) says: 'The major environmental
organizations now command resources and use a range of tactics
similar to anything deployed by the traditional economic interests
on which most theories of interest group politics are based.'[22] The
sorts of explanations that are commonly associated with the environmental movement (values, organizational style) may be

restricted to the small direct action (ecological) fraction of the movement and may not be fairly generalized to the broad gauge sense of the environmental public. We, like Bosso (1991: 167), see profound cleavages within the environmentalist cause:

> To the Greens [i.e. in his sense direct action groups], the major environmental groups have become just another set of lobbyists, and those holding 'deep ecology' values are just as unlikely to find common ground with their more mainstream brethren as they are to have trust in corporate or government leaders. The difference is one of values: mainstream environmentalists see their roles as ones of competitors in interest group politics; Greens shun institutional approaches for a fundamental reconfiguration of social values and behavior. *For the mainstream groups the issues are political; for the Greens the issues revolve around lifestyle* [emphasis added].

As Bosso (1994: 47) concluded, the environmental movement can no longer be seen as a single, identifiable unit if it is defined as comprising all 'organizations concerned with the question of human use of nature'.

Amnesty International and Friends of the Earth

The project reported on here surveyed five organizations (it also covered NFU Scotland and Aberdeen Chamber of Commerce); this book, however, concentrates on FoE and Amnesty International (British Section). For the purpose of comparison we can draw on information gathered about the members of one of the other bodies, namely, the National Farmers' Union Countryside.[23]

FoE and Amnesty are active in different policy fields but there are a number of important links which make it intellectually fruitful to study both groups in tandem. These organization have similar political operating styles; both are successful large-scale protest businesses; and their memberships are drawn from the same *predisposed public* in both socio-demographic and attitudinal terms (see chapters 4 and 5). There is a considerable degree of overlap between FoE and Amnesty in terms of membership, indeed there is overlap of individuals in membership (see Table 4.11).[24]

Amnesty International[25]

Amnesty International was launched by Peter Benenson (with the assistance of Eric Baker, a prominent Quaker; Louis Blom-Cooper, a well known lawyer; David Astor, editor of *The Observer*; and Gerald Gardiner who later became Lord Chancellor) with a front page feature in *The Observer* (26 May 1961). Its inaugural campaign – APPEAL FOR AMNESTY, 1961 – had a clear *raison d'être*: 'to work impartially for the release of those imprisoned for their opinions, to seek for them a fair (and public) trial, to enlarge the right to asylum, to help political refugees find work, and to urge the creation of effective international machinery to guarantee freedom of opinion' (Power, 1981: 10; Ennals, 1982: 65).

Six weeks after the launch it held its first international meeting which was attended by representatives from Belgium, France, Ireland, Switzerland, the United Kingdom, the United States, and the Federal Republic of Germany. By the end of 1961 there were Amnesty groups in Belgium, Greece, Australia, Sweden, Norway, Switzerland, France, West Germany, Ireland, the Netherlands, Britain, and the United States and the organization became known as Amnesty International (Power, 1981: 12). In 1961 Amnesty had 19 staff and an annual budget of £35,000; by 1981 its staff level had risen to 150, its budget to £2 million and it had a world-wide membership over 250,000 (Power, 1981: 12). In 1992 Amnesty International had approximately 1.1 million members, subscribers and regular donors in 150 countries, over 8,000 local Amnesty groups in 70 countries, and a budget of £12.75 million (Amnesty International, 1993).

Ennals (1982: 78) argues that: 'Amnesty International is founded on the very simple precept that governments respond to public opinion.' In deciding on its strategies Amnesty, according to Ennals (1982: 79) should have one paramount consideration: '.. what will be the most beneficial to the interests of the prisoners involved? It may be that persuasion by friends is more effective than publicity by those thought to be hostile. It may be that publicity on a massive scale is the only possible way of gaining attention and rights.' Ennals (1982: 65) points out that: 'The core of Amnesty International is the groups.' *Local Campaign Groups*, *Adoption Groups*, and *Affiliate Groups* located in religious or educational establishments, or trade unions can be any size, comprising a network of individuals who are committed to work for the release of *prisoners of conscience*. Local group members commit time (e.g. letter writing and other campaigning activities) and financial resources (e.g. membership subscriptions) to Amnesty International.

The British Section of Amnesty International is composed of three discrete entities: Amnesty International British Section (unincorporated), which organizes the campaigns and is responsible for fund-raising and administration; Amnesty International British Section Limited, which owns the properties and is involved in certain trading activities; and Amnesty International British Section Charitable Trust which is responsible for promoting research and providing relief to victims of human rights abuses.

Finance

Amnesty International's main source of finance is membership dues, and donor/trust income. In 1992 membership dues accounted for £1.5 million and donor/trust income for £1.1 million from an annual income of £3.92 million (see Figure 1.2). There are strict conditions on the sources from which Amnesty International will accept monies. The restrictions aim to protect the autonomy of the organization from anyone trying to buy favour. For example, no group or section should be dependent on one single source of funds. Some national groups can benefit from tax exemption (e.g. Amnesty International British Section Charitable Trust). Figure 1.3 shows the pattern of Amnesty's expenditure for the years 1990–94.

Internal democracy

Membership control of Amnesty International is by *National Sections* which send a representative to an annual *International Council Meeting* where the *International Executive Committee* is elected. The Committee directs the work of the International Secretariat and appoints a Secretary General who is accountable to the Committee for the Secretariat's activities and the implementation of Council Directives (Ennals, 1982: 66). There is also a national level conference.

Ennals (1982: 70) argues that democratic control via the decision-making process 'is a more difficult area in which to stimulate or maintain effective consultation, *let alone control*' (emphasis added). Ennals (1982: 70) further argues that the decisions which Amnesty takes over its priorities, tactics, and strategies are to a certain extent dictated by the vagaries of international events. These decisions are taken at the headquarters and are 'largely a secretariat

Figure 1.2 *Sources of Amnesty International British Section income, 1990–1994 (£ millions)*

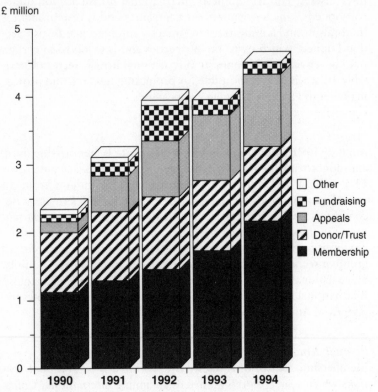

Source: Amnesty International (British Section), 1993: 14–15.

area of responsibility under the supervision of and in consultation with, the elected International Executive Committee'.

Within the British Section members of Amnesty International are entitled to vote at the Annual General Meeting (AGM), as are Local Campaign Groups (five votes), Adoption Groups (ten votes), and the number of votes that an Affiliate is entitled to is dependent upon the size of the group. At the AGM resolutions can be submitted by individual members or groups and any resolutions which are passed become Amnesty International British Section policy (Amnesty International British Section, 1993: 17). The Council is responsible

Figure 1.3 *Pattern of Amnesty International British Section expenditure, 1990–1994 (£ millions)*

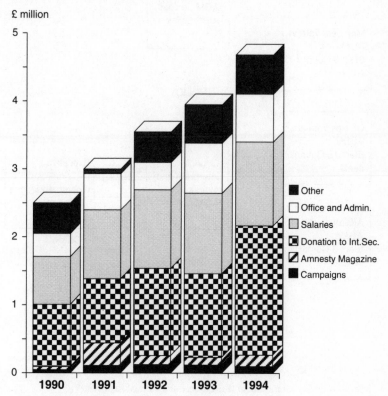

Source: Amnesty International (British Section), 1993: 14–15.

for ensuring that AGM policy is implemented and makes decisions on major policy issues which fall between AGMs. The Council is composed of five individual members, fifteen group and three affiliate delegates, two specialist group and five country specialist group members (Amnesty International British Section, 1993: 17) (see Figure 1.4).

Though there is a formal democracy with an AGM, attendance is inevitably low (as a percentage of support) and the normal decision-making process within Amnesty is within the Secretariat. Ennals (1982: 70) states that it is:

Figure 1.4 *The structure of Amnesty International British Section*

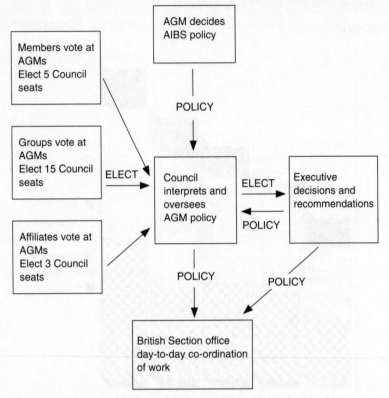

Source: Amnesty International (British Section), Annual Report 1993.

difficult to involve the membership in strategy regarding individual countries. Attempts are made to plan strategy by calling regional meetings but
the tendency is for the influence in such consultative processes to rest with
the International Secretariat because it is they who have the professional
experience ... Amendments to the statute which relate solely to the very
purposes and scope of the organization must be the sole responsibility of
the membership expressed in and by the Council. The International Secretariat and the Committee take part in these debates as part of the movement
but the decision itself rests with the membership. On the other hand, the
major day-to-day decisions must remain with the Secretariat which decides
which prisoners of conscience are to be adopted and which cases require
further investigation because no decision can be taken immediately on
whether they fall within the terms of the statute.

Friends of the Earth[26]

The first FoE Inc. organization to be founded was in San Francisco on 10 July 1969 by David Brower. He was the former executive director of the Sierra Club and he was *effectively* 'removed from that post (via an election defeat) because his confrontational methods had led to the Sierra Club's loss of its tax-exempt status' (McCormick, 1991: 33; see also Mundo, 1992: 173). Almost immediately after departing from the Sierra Club, Brower decided to found a new organization dedicated to the specific task of 'waging political battles to protect the environment' (Burke, 1982: 105). (Needless to say, Brower did not seek tax-exempt status for the new organization.)

FoE was always intended to be an international organization and in January and June 1971 delegates from six countries – France, the Federal Republic of Germany, Italy, Switzerland, the United Kingdom and the United States – met and devised a set of Governing Provisions. A Council of Friends of the Earth International was set up as the governing authority with the power to admit and suspend members (Burke, 1982: 107). In 1991 there were FoE (International) groups in forty-four countries. These groups are autonomous but share the same goals and objectives.

Over many years FoE has developed a systematic approach to mobilizing to protect the environment. This has several guiding principles:

1   *To remain politically non-partisan* – FoE aims to influence whichever political party is in power, but does not have a close relationship with any particular party.
2   *To seek the prevention rather than the cure for environmental problems* – FoE's efforts are directed towards a holistic conception of environmental problems.
3   *To attack specific key issues* – FoE identifies several key areas and concentrates its activities on them.
4   *To build a strong full-time staff* – from its inception FoE has devoted a considerable degree of its time and resources to building up a large staff, as it believes that without a large professional staff it would have great difficulty influencing policy-makers.
5   *To argue from information rather than ideals* – everybody agrees with the importance of improving the environment; however, FoE believes that a concise and accurate amount of information is likely to achieve more than a tirade of idealistic rhetoric.

6   *To build a wide network of autonomous local groups* – FoE hopes to
    alter public opinion and it sees the permeation of knowledge via the
    local groups as an effective way to change the climate of opinion and
    consequently to change public policy (based on Burke, 1982: 111–12).

Finance

Friends of the Earth Limited was formally incorporated in the
United Kingdom on 5 May 1971 (Burke, 1982: 109).[27] FoE Limited
is effectively two separate organizations – FoE Limited (Registration

Figure 1.5 *Sources of FoE Trust Limited income, 1993–1994*

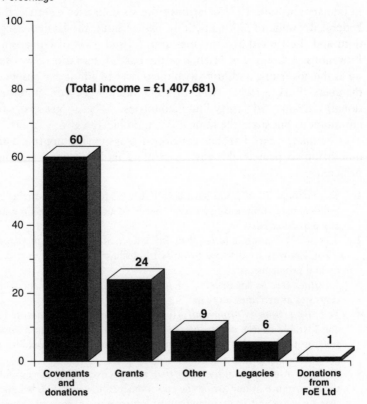

Source: FoE (1995), *Friends of the Earth Annual Review 1994* (London: FoE).

No. 1012357) and FoE Trust (Registration No. 1533942: Charity Registration No. 281681).[28] FoE Limited is 'a limited company by guarantee having as subscribers the original founders' (Burke, 1982: 109).

In 1971 FoE in London had 6 full-time staff, an annual income of £10,000 and 1,000 registered supporters (Lowe and Goyder, 1983: 133) and most of its money was spent on staff costs and publications. In 1994 it had 96 staff, an annual income of £5.3 million, and 112,000 supporters.

Figure 1.6 *FoE Trust Limited expenditure, 1993–1994*

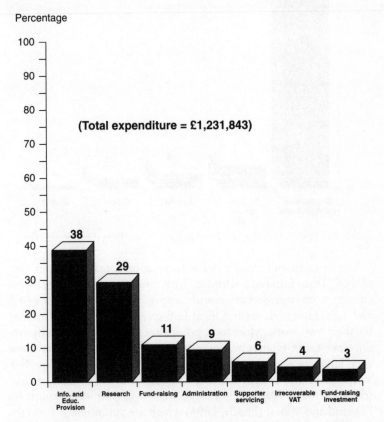

Source: FoE (1995), *Friends of the Earth Annual Review 1994* (London: FoE).

Figure 1.7 *Sources of FoE Limited income, 1993–1994.*

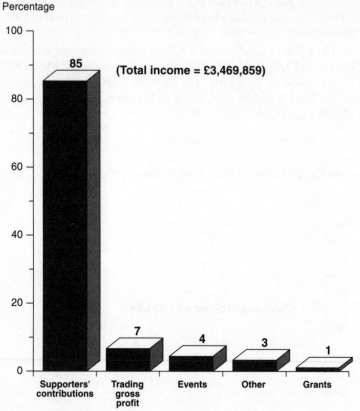

Percentage

Source: FoE (1995), *Friends of the Earth Annual Review 1994* (London: FoE).

Membership and trading account for approximately 65 per cent of FoE Trust Limited's income, with the remaining 35 per cent coming from large donations and campaign appeals (see Figures 1.5 and 1.6). However, several local FoE groups raise sums of money for their own work. Most local groups are small and are run by volunteers raising relatively modest sums. However, there are some very large local groups. For example in 1982, Birmingham FoE 'employed thirty people and had a budget of over £150,000 a year' (Burke, 1982: 110). In 1994 there were over 300 local groups (in England and Wales) (Frisch, 1994) which are autonomous from the London-based organization and which campaign on issues of their

Figure 1.8 *FoE Limited expenditure, 1993–1994*

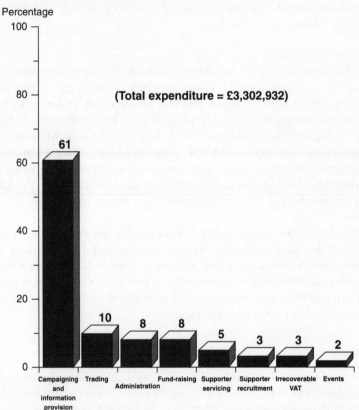

Source: FoE (1995), *Friends of the Earth Annual Review 1994* (London: FoE).

own choosing. Not all FoE 'supporters' are members of local FoE groups and the local groups are not required to contribute financially to the national organization. Since each of the local groups is independent from FoE Limited, their income is not included in FoE Limited's accounts (see Figures 1.7 and 1.8).

FoE in the United Kingdom relies more heavily on supporter contributions than the average environmental group in the United States. It also receives considerably less in the way of foundation grants and other patronage benefits (cf. Figure 1.5 and Figure 1.1). FoE in Britain has now changed its policy on accepting gifts/donations from business organizations. It now actively seeks sponsorship via its Corporate Fundraising Director. However, FoE acceptance of funds is 'conditional on companies' stance on ethical and green issues' (Rawcliffe, 1992: 4).

Internal democracy

The organization's subscribers appoint the Board of Directors to run the company and the Board appoints the staff. According to Burke (1982: 109) the supporters play no direct role in forming the policy of the organization. (This is seen as an important issue in the next chapter when the issue of FoE as a (New) Social Movement Organization is discussed.) FoE Limited admits members at the Directors' discretion. Individuals sending in annual contributions are constitutionally and technically 'Friends of the Earth'. A 'Friend of the Earth' is *de facto* a supporter of, or donor to, the organization.

Under United Kingdom law FoE Limited is not entitled to claim charitable status because of its political objectives. FoE Trust is a charitable organization which is entitled to apply for support from funding bodies which support charities. FoE Trust's charitable status restricts it from becoming involved in political campaigning and it is involved in environmental research.

Notes

1   The environmental issue appears to be a fairly contemporary phenomenon. The word 'environment' did not appear in the General Election manifestos of the main political parties in the United Kingdom until 1970. However, the Commons, Open Spaces and Footpaths Preservation Society was founded in 1865 and is reputedly the oldest national environmental organization in the United Kingdom. In the United States the Sierra Club was founded in 1892 (Cotgrove, 1982: 12).

2   It is questionable whether the National Trust should be included as the same type of environmental organization as the RSPB, WWF, Greenpeace, or FoE, because a substantial proportion of its membership have joined solely for the selective benefit of free entry to the Trust's properties, rather than as a statement of support for the 'other regarding' environmental causes.

3   The MORI questionnaire contained 369 questions and a sample of 1,458 respondents in 251 parliamentary constituencies in England, Scotland and Wales between 2 and 13 March 1989 (Jacobs and Worcester, 1990: 10).

4   Only 45 per cent of respondents agreed with this statement in 1981. However, despite the growth in support for environmentalism among the general public, as Bosso (1991: 157) points out: 'Environmental values are not yet as deeply ingrained as environmentalists might like ... Society is still marked by a throw-away consumer culture and a (poorly challenged) perception that there is a clear trade-off between a clean environment and a healthy economy.' There may be a significant disjuncture between attitudes and behaviour. Individuals' attitudes may have changed, but their behavioural patterns remain largely the same. People may know what constitutes environmentally friendly behaviour, but they may not conduct their daily lives according to such mores. In addition, there can be little dispute that many people believe that 'the environment' is a 'good cause'. Such 'apple pie' perceptions tend to elicit positive reactions – i.e., high positive ratings for environmental concern. To answer in the negative may reflect badly on self. Johnson (1995: 4) has pointed out that many studies (e.g. Scott and Willits, 1994; Stern *et al.*, 1993; and Van Liere and Dunlap, 1980) have found rising levels of environmental concern, but they have also 'found that widespread support for pro-environmental attitudes is not accompanied by pro environmental behavior; a relatively small number of people with strong collective concerns follow through on those attitudes by taking individual action'.

5   Costain and Lester (1995) cite annual column inches of references to environmental topics in the *New York Times* in different decades. The average per year increased steadily from 3.1 inches in the 1890s to 40.1 inches in the 1940s. There were then rapid increases to 110.3 inches in the 1950s and 284.5 inches in the 1960s and an even bigger leap to 944.7 inches in the 1970s (falling back in the 1980s).

6   Burke (1982: 112–113) argues that the history of the environmental movement in the United Kingdom can be divided into three broad phases. He sees the wildlife protection and preservation phase as the second phase, and the new wave of environmentalism as the third. He argues that the first wave of the environmental movement 'begins with the Levellers and the Diggers, draws inspiration from Kropotkin and the nineteenth-century anarchists, as well as from Robert Owen and the co-operative movement in its diverse forms. Today it manifests itself in the rural and urban co-operatives, the radical health and personal development groups, and the community of arts organizations, the self-sufficiency movement and many other forms ... its strongest characteristic is a commitment to act rather than argue about change'. He continues 'Within FoE the very many people making conscious efforts to live their lives according to ecological principles constitute our strongest bridge to this part of the environment movement.'

7   However, in the 1980s and 1990s environmental groups have

tended adopt a more 'positive' and 'pragmatic' approach.

8   By 1995 Greenpeace was conceding that its support had waned, but still claimed nearly half a million members and an income of £9 million.

9   The periods of growth (and decline) appear to be highly concentrated, supporting a fashion interpretation of membership. For example, McCormick (1991: 152) pointed out that: 'In the eighteen months from July 1988 to December 1989, membership of Friends of the Earth grew by more than 200 per cent, from 39,000 to 120,000.' Rawcliffe (1992: 3) notes that growth rates were particularly concentrated in the period 1987–1990. For example, in 1988, Greenpeace, 'gained some 130,000 members at a rate of over 4,000 a week; in total an increase of 68% from the previous year'. The ephemeral nature of the membership explosion is demonstrated by Greenpeace USA's loss of some 550,000 members between 1990 and 1992 (see Table 1.2). Bosso (1995: 103–4, 112) argued that the 1990s brought stagnation and/or decreases in membership and finance to environmental groups which lead to 'widespread retrenchment'. Staffing levels were cut in the Sierra Club, the Wilderness Society, and Greenpeace – Greenpeace laid off 50 of its 200 or so staff.

10   Greenpeace may be seen as occupying the 'purist's high-ground' but it doesn't stop it using professional (and successful) fund-raising techniques to fund its operation. For example, in 1994 Greenpeace in Germany received 71.2 million Marks in donations (*SPIEGEL Special*, November 1995).

11   However, while the number of those prepared to contribute has grown, a very small minority (around 5 per cent) of respondents claimed to have become actively involved.

12   In the United States the substantial incomes of many environmental groups have led these organizations to pursue their aims and objectives via a number of lobbying strategies. For example, the Sierra Club (akin to many other interest groups) established its own Political Action Committee (PAC) – Sierra Club Committee on Political Education (SCCOPE) – which spent $231,125 in 1981–82; $250,919 in 1983–84; $250,042 in 1985–86; and $286,904 in 1987–88 on Congressional elections (Mundo, 1992: 192). The Sierra Club also has close links with the formally separate organization the Sierra Club Legal Defence Fund (SCLDF). According to Mundo (1992: 194) the SCLDF pursues Sierra Club objectives through the courts: 'SCLDF officials consider the significance of a case in the light of the overall goals and strategies of the Sierra Club and the SCLDF ... The SCLDF and the Sierra Club operate on the basis of mutual approval with respect to decisions to litigate. The SCLDF cannot litigate without the Sierra's Club's approval, and the Sierra Club cannot force the SCLDF to take a case it does not want.'

13   The fact that these organizations have operating budgets akin to

small or medium sized companies has led to claims that they have 'sold out' and are no longer the true champions of public interest causes. Professionalism is seen as eroding idealism. This perception is given added weight by the fact that many of the more 'successful' organizations – in terms of membership and budgetary size – in the United States are paying their top executives over $100,000 per annum. In 1991, Jay Hair, President of the National Wildlife Federation earned approximately $300,000 (Bosso, 1995: 106). However, such comments may set up a false contrast between professionalism and sincerity.

14 For many public interest groups turnover rates of between 30 and 40 per cent are seen as 'relatively healthy' in this highly competitive market.

15 It should be noted that the data in Table 1.4 identifies a different membership pattern than the FoE entry in Table 1.1. This merely serves to illustrate that it is as difficult to get reliable membership data on the public interest group sector as it is on political parties.

16 Organizations such as FoE and Amnesty are seen as an important part of 'New Politics' which Jahn (1993: 177) defines as having 'four simultaneously emerging elements'. First, 'social issues such as environmental pollution'. Secondly, 'new modes of collective action'. Thirdly, the growing importance of post-material values. Fourthly, the participants in the new politics tend to be drawn from the young and the highly educated who are more likely to hold post-material values.

17 Earlier in 1995 Shelter, too, was the subject of controversy when the press disclosed that it spent more than £1 million purchasing its London headquarters, and that over £3.5 million of its £10 million income went on salaries and tens of thousands went on company cars (*Sunday Times*, 2 July 1995).

18 The analogy can be extended by seeing the various national forms as a sort of franchise operation.

19 In 1993/94 FoE Limited spent around £165,100 on supporter servicing, £99,000 on supporter recruitment, and £264,200 on fund-raising (estimates from Friends of the Earth *Annual Review*, 1994).

20 Rucht (1990: 169) identifies three main groupings within the environmental movement: *traditional conservationists* who perceive themselves as non-political and pursue their goals through responsible and moderate forms of action; *environmentalists* who use *conventional* 'pressure group' strategies and tactics in pursuit of their objectives; and *political ecologists* who are involved in disruptive, sometimes violent, grassroots participation.

21 Allan Thornton, the former Executive Director of Greenpeace United Kingdom, noted that: '... it moved away from "lifestyle" campaigns which concentrated on efforts to change public habits such as car driving. One of its most promising campaigns, on transport and car pollution was quietly closed down' (*Observer*, 10 December 1995). This was, of course, precisely against the trend sought by those with an ecological or social movement perspective. A reason might have been that attacking large unpopular targets like the French government or the nuclear industry is better 'box office' in terms of money-raising activities.

22   As Bosso (1991: 153) has highlighted the 1990 Earth Day in the United States was a professionally run event. It was headed by Denis Hayes, had more than two dozen paid professionals on a $3 million budget, had business and labour leaders on its board, used computers donated by Apple and Hewlett-Packard and extensive direct mail advertising, and products were licensed to raise income.

23   This is a money-generating idea by the NFU to recruit members for the Countryside operation with small recreational farms or even large gardens. This, it is hoped, can use the general expertise of the NFU and apply it to a different sub-population.

24   Our data, for example, showed that 12.8 per cent of FoE members are also members of Amnesty, while 16.9 per cent of Amnesty members are also members of FoE. It is also worth noting that 31.9 per cent of FoE and 33.7 per cent of Amnesty members were also members of Greenpeace.

25   This section draws heavily on Martin Ennals' 1982 essay – 'Amnesty International and Human Rights'. He was Secretary-General of Amnesty International from July 1968 to June 1980.

26   This section draws heavily on Tom Burke's essay – 'Friends of the Earth and the Conservation of Resources'. He is a former Vice-Chairman of FoE Britain.

27   The United Kingdom has two FoE International members – Friends of the Earth Limited and Friends of the Earth Scotland. FoE Scotland was formed as a separate legal organization in anticipation of devolution and in response to environmentalist sympathies for regional movements (Burke, 1982: 124). While FoE Scotland is a separate organization from the London-based group its national advertising means that the latter has members living in Scotland.

28   It is worth noting that not all environmental and campaigning organizations qualify as charities. *The Charities Digest* (1995, 101st edn: viii) states that: 'To qualify as a charity an organisation must be for the relief of poverty, or for the advancement of education, or for the advancement of religion, or for other purposes beneficial to the community, not falling under any of the preceding heads. The fourth category is the widest but cannot be more closely defined because the law recognises that the scope of charity alters over the years with social conditions and ideas. Legally charitable objects must benefit the public covered by one or other of the above heads (e.g. disabled people). The four heads enumerated above have been the subject matter of extensive judicial interpretation.'

2

Old political science and new social movements

Nothing is so beguiling in social science as the word 'new' ... the 'new functionalism', the 'new institutionalism', the 'new political economy', the 'new marxism', or, more recently, the 'new corporatism'. What the word 'new' really tells us, however, is that the new phenomena attacked by the new approach are not so new, after all, and that the new approach is a revival of some 'old' ways of seeing and dealing with things ... There is a good deal of revivalism in social science as there is in religion ... even though the new apostles are unwilling to acknowledge it, either out of ignorance or sheer perversity.

(Eulau, 1989: 19–20; cited in Kasse, 1990: 84)

Perspectives on environmentalism

There are two major sorts of discussion[1] about environmental participation: the *social movement perspective* and the *pressure group perspective*. The social movement perspective emphasises a *lack of hierarchy and formal organization*. Those active in politics, it is argued, now express their political frustration via these loose configurations rather than traditionally bureaucratic and hierarchical political organizations (e.g. political parties and formal interest groups). The social movement perspective sees the type of activity, and structures within which it takes place, as distinctive. The pressure group perspective focuses on the groups' relationship to the political system (Lowe and Goyder, 1983: 3). From Madison, to Bentley to Truman, many commentators have seen citizen formation of, and participation in, organized pressure/interest groups[2] as a healthy aspect of the political systems of democratic regimes. In this perspective groups enhance democracy by offering citizens a greater degree of participation in the political system.

This book mainly reports data collected through a survey of the members of three British organizations, two of which *can* be seen as part of the *new politics* associated with the rise of new social movements (NSMs)[3] – FoE and Amnesty. As explained earlier, we also surveyed an organization which is not closely associated with *new politics* – the NFU Countryside – which recruits 'country' people and may be seen as a part of the conservation wing of the environmental movement, but it is not a campaigning organization. The vast majority of FoE and Amnesty members are motivated to 'participate' to secure 'public goods' by a general sense of altruism, with there being a very limited role for selective incentives. On the other hand, the vast bulk of NFU Countryside members are motivated by the group's provision of selective material incentives (see chapter five).

This chapter discusses different academic treatments of groups such as FoE and Amnesty. Of what are these specimens examples? Our perspective is to treat these organizations primarily as *citizen groups* in the terminology of Walker (1991) and thus we operate primarily in the pressure/interest group perspective: this is the interpretation of mainstream political science (at least in the Anglo-American tradition). The essential quality of the citizen group is that membership is not based on occupation. Walker (1991: 61) noted that:

The citizen groups category includes almost all the public interest groups organized around ideas or causes ... Citizen groups are the kinds of groups that most sharply face the organizational dilemmas pointed out by Olson [see our chapter three]. Very often their fundamental purpose is to procure public goods such as world peace or clean air.

However, as implied above, such public interest groups have also been discussed in terms of a *social movement* literature. This concept has more sociological roots. It is this term – rather than interest group – that is probably the most commonly used to discuss the phenomenon of environmental support. What is the *added analytical value* of the social movement perspective?

In our view the social movement term is usually underdefined, which almost inevitably leads to difficulties in discerning the appropriateness of the interest group or social movement terms – and the different implications, if any, of these different paradigms. Indeed,

Eyerman and Jamison (1989: 1) say: 'In conceiving social movements ... sociology has all but made it impossible to understand them.' This is a conclusion we endorse. So while this book primarily treats multi-member (or at least supporter) environmental organizations as pressure/citizen groups, we recognize that (many) others deal with them as an integral part of a *social movement*. It is far from clear whether this reflects a real difference or simply a labelling choice. Definitional problems, as McAdam *et al.* also point out, are severely hampered by the diverse 'range of phenomena lumped together under the heading of social movements' – e.g. public interest lobbies, revolutionary and religious movements, etc. (cited in Lidskog, 1994).

The significance of terminology

The issue of which term, and associated conceptual framework, is important because terms carry implications with them. There is a danger of conflating material from these two discrete perspectives: generalizations of the characteristics identified as relevant for particular fragments of the so-called environmental movement (i.e. formally organized groups) may be inappropriately extrapolated to the broad whole. Or, alternatively, characteristics associated with a core of activists may be extended to cover the large numbers of individuals whose activism is restricted. The characteristics of the most active are assumed as applying to everyone who uses an ozone-friendly aerosol, or engages in green consumerism, or makes a financial contribution to the group.

Dalton's (1994: 3) book on environmental groups in Western Europe, *The Green Rainbow*, begins by claiming that: 'Environmental activists criticize the excesses of today's consumer-driven society and demand a restructuring of life-styles and the economic system.'[4] He further argues that the environmental movement shares similar characteristics (e.g. political style, organizational tendencies, even the personnel overlap) with other NSM (see below) – the women's movement, peace groups, and the self-help movement. 'Collectively, these movements are calling for a new agenda for contemporary societies and demanding that governments open the political process to more diverse and citizen-oriented interests' (Dalton, 1994: 5).

In our view this generalizes from the characteristics of a politicized and active part of the environmental population to a 'softer' but larger grouping. A description appropriate for the activists in 'Justice' (which published the *DIY Defence Manual* which gave advice for 'eco warriors' in 1995 about disabling road-building machinery and targeting construction firms) would hardly be accurate for the membership of the Council for the Protection of Rural England whose spokesman complained that such tactics gave 'legitimate forms of protest a bad name' (*Sunday Times*, 15 January 1995).

At issue then is the relationship between different 'environmental populations'. Those who are broadly sympathetic to environmental goals and might dutifully take their wine bottles to be recycled (but reject changes that would cause them to alter their own lifestyle to the extent of significantly reducing car use, foreign holidays, etc.), might be very different from those that are actually in membership of a formal organization such as FoE, and very different in other ways from those who have a well-developed collective identity on the issue. Within the faction actually mobilized into group membership there may be divisions between soft environmentalists holding what we can term *consumer membership* (making a very low investment in participation) from those ecological-type activists who have rigorous pro-environmental goals.

When the term '*social movement*' concept is used, does it mean the broad unorganized sea among whom are found the group members who have a greater commitment and express it by formal membership? From this perspective the environmental movement includes all those individuals who are concerned about pollution, nuclear power, the quality of life, and related issues. For example, Robinson (1992: 46) points out that many people see the environmental social movement as an 'open group', which means that:

anyone can be an environmentalist; it does not require one to join a group, or hold a definitive environmental ideology. The idea of environmentalism as an *organised* social movement hinges on the interpretation of formal core, consisting of identifiable fluctuating groups able to coalesce with other groups in given circumstances. However, part of the social movement idea is that it is deliberately unorganised, decentralised and diffuse in terms of its mass support. The important point is that environmentalism possesses the necessary spread of support from which it can potentially draw *when required*.

Alternatively, is the social movement a *cadre* idea that implies a level of commitment that is *stronger* than that expressed by mere group membership? To put it bluntly, who is part of the social movement: people putting lead free petrol/gas in their vehicles, or those who have changed their lifestyle – not just their carburettor? Some, perhaps optimistically, seem to imply that the big battalions of those contributing to mail-order protest businesses are the same as the ecologically committed.

Jamison *et al.* (1990: 186) say that the environmental movements they studied (in Sweden, Denmark, and the Netherlands) shared a common set of *knowledge interests*: 'The worldview of systems ecology provided environmental activists ... with a common frame of reference, a shared cosmology. According to the ecological world view all is interconnected: the local, the national, and the global'. Martell (1994: 188) cites Barho's rejection of Die Grünen in 1986 because it failed to support a total ban on animal experiments. Barho (1986) instead offers an agenda that is about spiritual renewal through the proliferation of green communes that encourages ecologically sensitive lifestyles. This kind of programme may be green, but is it a shade that will appeal to large numbers?

Contrary to the impression given by social movement interpreters we see such radical portrayals as appropriate to only a restrictive interpretation of 'activist'. One cannot assume that those 'active' only to the extent of supporting relevant campaigning organizations hold such world views. There is a mismatch here. There is a population with a deep, transformed, and fundamentally different political perspective on environmentalism, and there is a large and broad population active in large-scale environmental organizations. *But in our account, these are very different populations. If the support is deep it is narrow: if it is broad it is shallow.*

Lidskog (1994: 2) notes that there are four ways of legitimating one's own theory of social movements compared with rivals. These, he says, are ignorance, criticism, integration, and relativization. It is very difficult to summarize, far less contribute to, the social movement approach because of the common preference for ignorance as a strategy. This is not to say that the approach is wrong, but that it has to develop without reference to the other literature. It has not sought to demonstrate how it relates to other views, but it has ignored them. Lidskog notes that Touraine (1981), for instance, independently constructs his own theory, and to the extent that he takes other theories into consideration at all, it is for the most part a question of cursory references. There is then no sense of aggregation in the literature: more turns out to be more confusing rather than more convincing.

Within much of political science the tendency is simply to conflate the group/movement concepts or to see them having a difference in terms of development over time. The social movement, in this interpretation, is the *pre-organization*. Bosso (1994: 34) quotes Snow as follows: 'even the most grass-roots campaigns of mass mobilization, if they survive, tend eventually to seek their place upon the institutional bedrock. The process is familiar enough: *Ad hoc*, inchoate groups of volunteers harden into chartered organizations, which if successful, evolve over time into long-standing, stable institutions.' This evolutionary concept of the relationship between the movement and the group is popular, but at the same time fiercely opposed by those who see the two ideas as fundamentally divergent.

Resource mobilization versus identity interpretations

Cohen (1985) describes the social movement field as split into two distinct paradigms. There are 'identity theorists' such as Touraine and Habermas who conceptualize social movements in abstract terms and 'resource mobilization' (RM) theorists who define social movements more empirically and focus on groups' tactics and successes. The conceptual overlap between groups and movements has been particularly large where *resource mobilization theory* has been utilized (see McCarthy and Zald, 1977). As Dalton (1994: 6) observes: 'Among political scientists, many of the theoretical interests and predictions of the resource mobilization paradigm are paralleled by the rational-choice model of political groups.'[5]

The rational choice perspective has been at the heart of the recent debate in political science. It is dealt with more fully in chapter 3. In this perspective the emergence of organized entities from a general dissatisfaction 'is not primarily a reflection on the existence of political grievances in society: instead it depends on the presence of sufficient resources and entrepreneurial expertise to create and sustain a movement' (Dalton, 1994: 6). Olson (1971), and his famous *free-rider* concern, is accepted as relevant by those utilizing a resource mobilization approach – 'grievances or deprivation does not automatically or easily translate into social movement activity, especially high-risk social movement activity' (Zald; cited in Morris and Mueller, 1992: 6).

From old to new social movements: the consequences

The boundary problem between the interest group and the social movement ideas has been aggravated by the further distinction between 'traditional' and 'new' social movements. While it *may* be the case that 'old' social movements have a recognizable social base and reflect class-based conflict, NSMs are held to reflect lifestyle and consumption issues.[6] Commenting on Melucci's work, Keane and Mier argue that:

contemporary social movements are not preoccupied with struggles over the production and distribution of material goods and resources ... the constituent organizations of today's movements consider themselves more than instrumental for attaining political and social goals ... movements are only part-time participants in the public domain. They normally consist of 'invisible' networks of small groups submerged in everyday life.
(Keane and Mier, Introduction to Melucci, 1989: 5–6)

Dalton (1994) cites Offe's (1990) suggestion that NSMs such as the environmental movement are based on the values of personal integrity, recognition, and respect. While Martell (1994: 112) distinguishes between old and new movements along the lines of *location, ideology and aims, organization of movement, and medium of change* (see Table 2.1).

Table 2.1 *New and old social movements*[7]

	Old social movement	New social movement
Location	Polity	Civil society
Ideology and aims	Political integration and economic rights	Autonomy civil society, new values/lifestyles
Organization of movement	Formal and hierarchical	Informal network and grassroots
Medium of change	Participation in political institutions	Direct action and cultural politics

Source: Martell, 1994: 112.

In our interpretation, the campaigning organizations discussed in this volume are (perhaps surprisingly) located towards the older rather than the new form, though Martell (1994: 208) cites sources such as Dalton et al. (1990) and Offe (1985) that argue to the contrary. Martell follows Scott (1990) who (like the resource mobilization school) sees the new social movements as more likely to be potential 'old social movements' at a stage in their development than some new form.

Scott argues that the notion of *newness* of new social movements has been exaggerated. They are *de facto* closer to old social movements than many (largely sociological) commentators (in search of novelty) have been prepared to concede. First, 'modern' social movements cannot be seen as 'primarily social and not directly political in character' (Scott, 1990: 16), and secondly, they are not so exclusively confined within civil society.

Much of the social movement literature seems reluctant to see itself as a variant on the interest group theme, yet as noted above conventional pressure group accounts accept the social movement as an embryonic type of their species rather than a new phenomenon. Walker's *Mobilizing Interest Groups in America*, for example, has as its sub-title *Patrons, Professions and Social Movements* – the social movement is seen as a loose form of pre-interest group rather than a fundamentally different sort of concept. For example, Walker (1991: 187) argued that:

the second mode of political mobilization emerges from social movements that sweep through society from time to time, normally arising from the educated middle class. Movements to abolish slavery, prevent the production and sale of alcoholic beverages, shut off immigration, ensure civil rights, or protect the natural environment have long been a central feature of American political life.

Thus there are in part both disciplinary and 'continental' differences in terminology which allow considerable confusion. Scepticism about the social movement approach is more likely to come from political science and from the United States. Eyerman and Jamison (1989: 2–3) point out that: '... a gulf has emerged between American and European sociologists in the understanding of social movements. While most American sociologists tend to fall into either of ... two categories (*Resource Mobilization* and *Particularist*), in Europe the most common approach has been to analyze social movements as carriers of political projects, as historical actors'. Dalton (1993: 8) discusses citizen protest groups in the 1960s in the United States and equates them with 'movements'. He describes a flowering of citizen action groups and social movements concerned with environmentalism, women's rights, peace, consumerism, lifestyle choices, etc. He says that similar types of citizen interest groups began to appear in Europe in the 1970s where they were known as 'new social movements'. He went on: 'We emphasize the groups and activities identified as new social movements in the European context or more loosely labelled as New Left-oriented "public interest groups" in the United States' (1993: 9).

'New' social movements are inconsistent with resource mobilization expectations; and so there often appears as much difference between different social/new social movement accounts as between them and political science/group discussions. However, there are at least five ways in which there have been attempts to distinguish social movements (both variants) from pressure groups: *degree of organization; clientele; method of influence; nature of goals*; and *ideological bond.*

The distinctiveness of new social movements?

Degree of organization?

As a broad generalization the social movement approach(es) deal with less rather than more formal and organized forms of political participation. These movements are said to prefer a decentralized, open, and democratic structure that is more in tune with the participatory tendencies of their supporters' (Dalton *et al.*, 1990: 13). (The anti-road building group Justice calls itself 'a **disorganization**' (*Sunday Times*, 15 January 1995).) That this sort of activity is *not* a part of the interest group world seems to rest on an *assumption* that interest groups are more formally organized than social movements. Thus, those who argue for special movement distinctiveness in terms of informal organization tend to be reacting against an extreme sort of stereotypical interpretation of interest group, rather than actual examples from the political science field.

The distinction between political science 'groups' and sociological 'social movements' is further eroded if relatively organized (institutionalized) variants of the social movement are identified.

Dalton records (1994: 6) how the resource mobilization (RM) theory focuses on *social movement organizations* (SMOs) rather than the nebulous movement. McCarthy and Zald (1973: 20) identified the development of *professional social movement organizations* (PSMOs). These take us further towards the mainstream of interest group studies. Common Cause is seen as a hybrid-type with its centre mirroring the PSMO form. ASH (Action on Smoking and Health) is seen as the 'prototype of the PSMO form' (McAdam, 1988: 717).

Dalton (1994: 7) points out that in the RM perspective the resource needs of SMOs influence their internal structure. Thus 'SMOs are inclined to adopt a hierarchical and highly routinized structure to maximize their efficiency in collecting money, activating members, and achieving policy success.' In an argument mirroring that conducted in the political science world, which stressed the pivotal role of the political entrepreneur (e.g. Salisbury, 1969), Dalton (1994: 7) argues that many SMOs would not exist without the initiative of a single individual or a small group of people. The RM perspective assumes that a requirement for political influence is a change in organizational form that takes SMOs nearer the stereotypical interest group. This approach – mainly in the RM mode – sees an almost inevitable process of development in which amorphous and fluid social movements become centralized and formalized. In this light the:

formally structured movement organizations are more typical of modern social movements and more effective at mobilizing resources and conducting sustained political campaigns … The professionalized and centrally organized group is presumably more effective in mounting a direct-mail campaign, in organizing a membership drive, in targeting its political activities on a specific goal, and in performing the other operational functions of a movement. (Dalton, 1994: 100)

Tilley (1978; cited in Dalton, 1994: 51) sees the SMO as the organized and developed expression of the views held by a broader supportive public. The supportive public legitimates the claims of the organization. Dalton accepts that the framework is 'ill defined and analyzed' but it does not share the RM expectation that effective social movements will be centralized and hierarchically structured organizations: 'While the resource mobilization literature treats this as dysfunctional and harmful to the effectiveness of the movement, the NSM literature maintains that this style actually serves as an attractive force for potential movement supporters' (Dalton, 1994: 9).

Dalton (1994: 101) presents NSMs as disconfirming the 'Weber-Michels imperative'. He cites Die Grünen (the German Green Party) and the European peace movement as the prime examples of the anti-organizational tendencies. He says that these NSMs will resist close ties to established political alliances which the RM approach sees as the path towards organizational effectiveness. He further argued that Die Grünen militants display an almost pathological aversion to oligarchy – as do the peace movement and the women's movement. But he also concedes that 'the actions of the party (Die Grünen) often fall short of its rhetoric' (Dalton, 1994: 9).

Undoubtedly the RM interpretation with its expectation of organization as the corollary of maturity and success is less distinct from the interest group literature. Moreover, the interest group approach has been open to include comparatively weakly organized activity. The NSM cases look less exceptional when studied in a context that accepts that so-called interest groups are often far less bureaucratized than is assumed by those looking for sharp contrasts. In fact, the NSM term is often used as a *mark of approval of the (radical) goal* rather than a statement about organizational structures that usefully distinguishes the group and the movement. As Morris says: 'Social-scientists whose accounts emphasize disorganization and spontaneity miss the mark by mistaking an image, projected to and taken up by repressive authorities, for the reality' (Morris, 1984: 75; quoted in Scott, 1990: 31–2).

Clientele

Lowe and Goyder (1983) see the environmental social movement as comprising two interrelated components: 'the organizational embodiment of the movement' and 'the attentive public'. Robinson (1992: 36) points out that the *organizational embodiment* is identifiable by its 'concise core beliefs and particular principles of action', and the *attentive public* is held together by a 'sense of communal purpose and ideology'. This construction means that one can be a member of the environmental social movement (i.e. the attentive public) without being a fee-paying member/supporter of groups such as FoE (i.e. the organizational embodiment). But it is possible to regard organized members of campaigning groups as less thoroughly informed and politizised than the so-called attentive public. Robinson (1992: 41) argues that:

Despite conceptual problems, one can consider environmentalism as being a coalition of groups with a singular, intrinsic ideological link ... However, it would be misleading to see the environmental movement as merely an amalgamation of *environmental* pressure groups. The terms of reference for the movement are broad enough to encompass cells of other groups whose ideological motivations are not derived from the sole perspective of environmental concern.

Burstein (1995) notes that according to some sources social movement organizations represent a particular 'market' – those outside established political institutions. (He cites McCarthy and Zald, 1977: 1217; Freeman, 1975: 45–47; McAdam, 1982: 25; and Tilley, 1984: 306.) As Burstein argues this seems unconvincing. If SMOs represent outsiders, then once they begin to succeed do they cease to be social movement organizations – though their goals, membership, or tactics stay the same? In fact, our data found that the mobilized members of the main public interest groups were more middle class, better educated, more likely to vote, and more likely to be members of a political party than the general public.

Scott (1990: 135) argues that social movements 'articulate the grievances and demands' of two main groups: 'groups who are excluded from the benefits typically available to average citizens' and those 'excluded from established elite groupings and from processes of elite negotiation'. Given the socio-demographic profile of the 'typical' member – educated (new/)middle class – this seems slightly surprising. This is not the group which immediately springs to mind when one thinks of the politically marginal and/or excluded. Scott (1990: 139) extends his critique of NSMs to attack the notion that these phenomena are not class-based movements. He claims that new social movements are class movements in Gellner's extended Weberian sense, and he further argues that the class based nature of these movements has, in fact, been recognized by authors such as Offe (1987) who says that: 'New middle-class politics is, in contrast to most working-class politics as well as old middle-class politics, typically a politics of a class, but not on behalf of a class' (Offe, 1987: 77, quoted in Scott, 1990: 139). Scott (1990: 139) concludes however, that Offe's distinction is shaky: '...

not because the new politics is reducible to class interest, but because the relationship between class interest and political practice is probably more complicated in all the cases Offe mentions'.

Method of influence?

Another dimension where the group and the movement have been distinguished concerns the method used to effect political change. Smelser (1962) claims that pressure groups seek to achieve change through traditional interest group intermediation patterns, while social movements are collective organizations which aim to change the social order through uninstitutionalized methods. As Dalton *et al.* (1990: 14) point out: 'NSMs seemingly prefer to influence policy through political pressure and the weight of public opinion, rather than becoming directly involved in conventional politics.'

Knoke and Wisely (1990: 57–8) claim that social movements (with political goals) are characterized by three main elements that included:

1 Socially disruptive actions targeted against public authorities and their symbols – ranging from peaceful public assemblies to civil disobedience and illegal violence – that transgress conventional norms of participation in electoral and pressure group politics.
2 Primarily purposive tactics and strategies, rather than emotional and expressive outbursts, that seek to shift power and policies from the status quo to new arrangements.
3 An emphasis on social organization that stresses a high degree of group activity rather than elite leadership.

Dalton (1994: 9) suggests that NSMs are different from social movements and interest groups because they reject neo-corporatist strategies of 'insider' political influence: many environmental organizations consciously reject neo-corporatism as incompatible with their goals and political style. Offe (1985: 830) also argues that NSMs are distinguished by a particular mode of action: 'This mode of action also emphasises the principled and non-negotiable nature of concerns ... Such a logic of thresholds, obviously, hardly allows for practices of political exchange or gradualist tactics. Movements are incapable of negotiating because they do not have anything to offer in return for concessions made to their demands.'[8] These anti-statist orientations presumably lead NSMs to adopt unconventional methods of political action (protests, demonstrations, and spectacular actions) as part of their political repertoire.[9]

In fact, such a proposition only distinguishes NSMs from *some* types of interest group. Many (so-called) interest groups have been recorded using such unconventional means. Dalton (1994: 256) asserts that the variety of organizational styles in the environmental movement makes it different from the rest of European interest group politics where most European unions, business organizations, and other interest associations follow the neo-corporatist model of unifying diverse interests under an umbrella group. This contrast can easily be exaggerated by over-estimating the effectiveness of umbrella arrangements.

Burstein (1995: 11) is sceptical of the 'method of influence' as a theoretically defining characteristic. Following Walker (1991) he argues that the use of demonstrations and 'unconventional' tactics is an attempt to reach potential members who are widely scattered through the population. Political demands have to be highly visible if they are to reach their target audience. From this perspective, public protest is simply a tactic useful for overcoming organizational problems, not a defining characteristic of a theoretically distinct type of organization.

Rochon and Mazmanian (1993: 87) argue that the relative success of the environmental movement was the result of its involvement in the policy process: 'The success of environmental movement organizations in affecting policy by gaining access to the policy process is a strategy that should be examined by other movements.' This perspective sees the social movement as more likely to be successful the less distinctive it is from an interest group. (But they note the problem of being outflanked by more militant organizations that do not take part in the bargaining process as a source of instability.) Gelb (1989) maintains that it is the 'political opportunity structure' which is 'a major determining factor of the particular style and significance of movement action' (quoted in Kuechler and Dalton, 1990: 288). Thus, while these movements are *forced* to mimic the strategies and tactics of institutionalized or conventional actors (e.g. political parties, interest groups etc.) because of the 'political opportunity structure', they should still be distinguished from them.

However, we prefer to follow Scott (1990: 132) who maintains that:

Contrary to culturalist interpretations, no categorical distinction can be drawn between social movements, pressure groups and parties. Social movements are best understood in terms of a continuum stretching from informal network-like associations to formal party-like organizations. We can realistically assess the effects of social movements upon their environment, by viewing them as a political phenomenon related to other more 'institutional' expressions of political interest.

NSM theorists argue that social movement activists favour 'grass-roots models of political participation' and 'unconventional and confrontational techniques' rather than the opportunities of representative democracy. But, as argued above, such unconventional participation has also been the hallmark of organizations such as Greenpeace – that are, elsewhere, included in the interest group category. Should Greenpeace be seen as a social movement, a social movement organization, or, a social movement interest group, or just a *plain* interest group?

Rucht (1990: 162) identifies two ideal types of movement strategy: a *power-oriented* strategy 'which is concerned with the outcomes of political decision-making and/or with the distribution of political power', and an *identity-oriented* strategy 'which focuses on cultural codes, role behaviour, self-fulfilment, personal identity, authenticity, *etc*. Such a strategy relies mainly on expressive behaviour, trying to change cultural codes by alternative life-styles'. Following from these *general strategies* there are several more *specific strategies* .[10]

Table 2.2 *Typology of social movements' strategies*

Logic of action	General strategy	Specific strategy
Instrumental	Power-oriented	Political participation bargaining; pressure; political confrontation
Expressive	Identity-oriented	Reformist divergence; subcultural retreatment; countercultural challenge

Source: Rucht (1990: 163).

Rucht (1990: 169) maintains that the 'contemporary environmental movement has been characterised as power-oriented'; although a small number of identity-oriented groups exist, they 'do not play a central role' in the movement. Within the environmental movement he identifies three main groupings – *traditional conservationists, environmentalists* and *political ecologists*. Conservationists tend to perceive themselves are non-political and pursue responsible, 'moderate and non-conflictual forms of action'. Environmentalists and ecologists are consciously and deliberately engaged in political activities, with the former engaged in conventional 'pressure group' activities of influencing the media and lobbying, and the latter involved in 'disruptive (and sometimes violent) means of action'. In general, conservationist bodies tend to be hierarchical while ecologists rely more on 'grassroots' participation. However, Rucht (1990: 170) points out that there are exceptions to this rule. He claims Greenpeace is based on an extremely rigid hierarchy, whereas 'Robin Wood', a German offshoot from Greenpeace, pursues similar goals and forms of actions based on a grassroots organization. However, as Peter Gundelach (1995: 438) has argued: '... even if grass-roots activity was protest activity in the late 1960s and 1970s, it soon became a standard part of the repertory of political activity. Political systems have not been seriously threatened by the mushrooming of grass-roots activity and new social movements.'

Nature of goal?

Some – particularly those writing on new social movements – focus on *goals* rather than organizational form. Zald and Ash (1966: 464) maintain that a social movement 'is a purposive and collective attempt of a number of people to change individuals or societal institutions and structures'. Social movements may have 'bureaucratic features' but these 'differ from "full blown" bureaucratic organizations in two ways': first, their goals aim to change 'society and its members'; second, the 'incentive structure' is heavily reliant on *purposive incentives* (value fulfilment). However (again), this sort of distinction only distinguishes the social movement from the sectional group; the focus on collective or sectional goals does not distinguish it from the promotional or cause group that has been widely identified in the political science literature.

Kitschelt (1993: 14) presents social movements as predominantly organized around the procurement of pure public goods rather than economic-distributive issues.[11] He says that political parties and established interest groups have been primarily organized around economic-distributive issues, that is, the allocation of scarce monetary resources to political constituencies. He identifies two foci in NSMs: *the politics of space* – environmental protection, land use, waste disposal and the like; and *the politics of social identity* – women's movements, abortion, anti-pornography, sexual harassment, gays, ethnics, handicapped, etc. Both currents oppose the commodification of society in a capitalist market economy. According to Kitschelt (1993: 15): 'This political thrust is reinforced by the experience that it is precisely the highly bureaucratized political party machines and interest groups ... that have been unresponsive to the demands of the left and libertarian social movement sector.'

The perception of social movements as loose structures in pursuit of libertarian collective goals is widespread, but that implies that interest groups cannot be libertarian or collective in their goals. For the sake of presentational symmetry it also assumes that interest groups are bureaucratic and social movements are not. In fact, so-called interest groups (traditionally defined) display a wide range of organizational forms (some of which are far from bureaucratic) and a wide range of goals.[12]

Moreover, the thrust of this kind of account is that the 'new social movement' is a political form adopted by 'the good guys'. But the expression of extreme nationalist, racist, or anti-abortionist views can take loosely organized forms with no direct selective benefit to participants: can one have a racist or an anti-abortion NSM? If not, is it because the term is only used for causes endorsed by the analysts, or is there some other way in which the use is to be restricted? Indeed, as Eyerman and Jamison (1989: 25) highlight: '... the sociologist, while using apparently neutral, even positivist methods in his analysis, becomes an apologist for the movement or movement event she studies'.

Kitschelt (1993:27) sees social movement informality as relating to particular decision points – abortion, equal rights amendments, etc. – that favour 'fluid direct democratic practices ... Such groups tend to develop a communal infrastructure of bookshops, coffee-houses, media projects ... that only intermittently engage(s) in political activity'. On the other hand, he says that feminist groups concerned with women's employment issues that require constant involvement in policy formulation and monitoring of implementation have developed formal interest group structures or have established themselves inside parties and interest groups. Kitschelt (1993: 27) argues that environmental tendencies are more likely to result in formal interest groups than is the women's movement because in the politics of space questions of self-organization, deliberation, consensus, and preference change are held important, but not essential to the attainment of the movement's objectives. He says that, in contrast, for the politics of social identity means and ends cannot be clearly differentiated. The proposition appears to offer a neat means to predict whether a policy area will generate more or less formally organized political structures, but unfortunately the connections are far less unambiguous than he suggests. Abortion has often generated well-structured organizations; the peace movement has generated structured and unstructured examples. Why should women's employment be better served by formal interest groups than is the abuse of women?

Ideological bond

As Lowe and Rüdig (1986: 515) highlight, it was Inglehart who was the first to identify the (post-material) value change with:

> the widespread shift away from the unquestioned predominance of economic and basic security values towards an increasing emphasis upon what Maslow had characterized as 'higher order' needs (i.e. the need for love, esteem and status and for intellectual and aesthetic satisfaction). The former Inglehart termed 'materialist' and the latter 'post-acquisitive', and later 'post-materialist' values.

The value change argument maintains that as a growing proportion of the public begins to emphasize post-material values the composition of the political agenda shifts from traditional economic and security concerns to the non-economic and quality-of-life values of a post-industrial society.

Kuechler and Dalton (1990: 278) define a social movement as 'a collectivity of people united by common belief (ideology)' which is shared by 'core members' and represents their *ideological bond*: 'the ideological bond between the core members determine the nature of the movement. It provides the prime criterion in determining whether the qualifier "new" is theoretically meaningful.' NSMs advocate a 'new social paradigm' which stresses post-material as opposed to material lifestyles, and the enhancement of participatory/direct democracy. The (one) *ideological* bond offers a 'humanistic critique of the prevailing system and the dominant culture ... and a resolve to fight for a better world here and now with little, if any, inclination to escape into some spiritual refuge' (Kuechler and Dalton, 1990: 280). The ideological bond is not, however, a 'strict doctrine':

The new movements do not follow some grandiose plan for a better society; they do not adhere to some Marxist (or any other) vision. Their concept of the future is largely negatively defined. They know what they do not want, but they are unsure and inconsistent with respect to what they want in operational detail. The absence of a strict doctrine may suggest that we could call these movements 'post-ideological' ... 'post-ideological' indicates the absence of strict doctrine and hierarchical organization rather than a lack of shared beliefs, the absence of an ideological bond. (Kuechler and Dalton, 1990: 281)

They further argue that the movements tend to shy away from violent tactics and most of the members are likely to be alienated from the dominant system. They claim that the coexistence of radical critique of the existing order, on the one hand, and the *de facto* integration into the existing society and into the political arena, on the other hand, is a genuine characteristic of the new movements.

In this less than unambiguous perspective what make NSMs 'new' and 'distinctive' is that they are characterized by alienation and integration, a radical critique pursued through 'unconventional' methods by a largely middle-class constituency materially benefiting from the existing order, and an negativist ideology which knows what it doesn't want, but not really what it does. A definition from the 'slippery when wet' dictionary.

Novelty or loss of professional memory?

An underlying concern in trying to resolve the real differences between these ideas is that the term 'new social movement' appears novel only if the history of, and relationships between, ideas is greatly simplified to allow 'space' for the new term. With regard to the issue of NSMs *newness* Scott (1990: 153) maintains that they are characterized by 'advocates' as reflecting: societal transformations such as the move from an industrial to a post-industrial society (Touraine) or from liberal to late capitalism (Habermas); the non-negotiable/non-accommodatable principles of their demands within industrial society; loose organizational structure; and belief in the advancement of basisdemokratie rather than outputs of the political system. Scott (1990: 154) does not take issue with the accuracy of these characterizations, but *whether such characterizations are exclusive to NSMs*. Such definitions are, he says, plausible only as long as one equates the movement as a whole with its fundamentalist wing for whom ideology is an ethic of absolute ends. Thus, in his view, none of the imputed characteristics is exclusive to NSMs. Enhancing and extending participatory democracy is applicable to social movements generally and 'is part of the rhetoric of populism' (Scott, 1990: 154). The evolutionary process which NSMs have undergone/are undergoing is, he asserts, analogous with that which the workers' movement has gone through – namely, the *fundamentalist/realist dilemma*. Once the initial 'revolutionary' calls have reverberated around the political system, how does the movement proceed? Does it remain ideologically pure and retain a non-negotiating stance, or does it adopt a more pragmatic position in the hope of attaining some of what it wants quickly, with the long-term aim of achieving the radical transformation via the politics of bargainable incrementalism?

This, according to Scott, is a dilemma faced by all movements once they are past their early stages of development. Scott (1990: 155–6) says that he is struck by NSMs' *ordinariness*. He concludes that:

This robs new social movements not of their significance, but much of their novelty. We cannot rest an entire theory of social transformation upon their presence, nor look to social movements alone as harbingers of some new society ... the search for some over-arching movement is fruitless given the diversity of demands and interests. This is a view perhaps disappointingly close to a 'common-sense' understanding. *But should a sociological interpretation of social movements be motivated exclusively by the search for novelty?* [emphasis added]

With regard to the environmental movement, the notion that the 'participants' are 'marginalized' or 'excluded' individuals trying to get access to influence appears less tenable. For example, several studies of membership of ecology bodies (with differing degrees of *organization*) – including our own – have shown that it is the middle classes who are the main participants – hardly the most economically, socially, or politically disadvantaged group in Western societies.

Zald and McCarthy (1987: 12) concede that their conception of an SMO is relatively encompassing 'sometimes coming perilously close to groups many would call "pressure groups"' (cited in Kuechler and Dalton, 1990: 278). In the light of the menu of definitions of interest or pressure groups set out below the question is whether the literature on social movements is a new field of study or if it simply represents a study of familiar subjects by those of a different ideological or intellectual persuasion. *Is the phenomenon of social movements more instructive of the politics of the profession than the politics of society?*

The wide attention given to social movements (as indicated in the range of papers, books, and articles that have appeared in recent years in contrast to the paucity on interest groups) has been imperfectly related to the interest group literature. It has assumed that interest groups are more formally organized with regularized membership arrangements than are in fact routinely covered in the range of interest group work. Undoubtedly the social movement literature is dealing with a material that is also covered by interest group writers – particularly those concentrating on cause groups or those who accept the concept of the unorganized group. As Burstein (1995: 5) argues, the differences between interest groups and SMOs is artificial:

I think it is fair to conclude ... that the distinction between 'interest groups' and 'social movement organizations', which initially seems obvious to those who study politics, *does not exist*, in the sense that no one has developed a convincing basis – theoretical or empirical – for distinguishing consistently between the two. Rather than continue trying to make the distinction, therefore, we should simply say that there are a variety of nonparty organizations which try to influence political outcomes; the organizations vary in a variety of ways of potential interest to social scientists (tactics, form or organization, number of members, resources, specific goals, etc.), but the simple dichotomy between 'interest group' and 'social movement organization' cannot stand up to scrutiny and should be abandoned.

He suggests that interests and movements should be studied together as *interest organizations*. Tonge (1994: 93) coined the term 'pressure movement' to cope with the fact that distinctions between pressure groups and social movements have become increasingly

arbitrary with the rise of groups lacking any formalized membership.

The social movement idea is also imperfectly distinct from a
political party. Bomberg draws attention to the phenomenon of the
'movement-party' such as Die Grünen. She argues that most literature on movement-parties sees an inevitable slide towards 'parlia-
mentarianism'. Consequently, she argues, movement-parties submit
to the normalizing forces of an established parliamentary system
and they begin to place increasingly more emphasis on parliamentary bargaining, compromise becomes more acceptable, and the
party itself becomes more preoccupied with maximizing votes
(Bomberg, 1993: 162). She cites von Beyme's claim (1984: 373)
that over the last century nearly all movement-parties have been
caught in the undertow of parliamentary government and have
ended up compromising and accommodating the imperatives of traditional party politics.

FoE: ecological movement, social movement organization, interest group, or protest business?

Dalton's *The Green Rainbow* focuses, like ourselves, on environmental interest groups, and offers a very different characterization
of FoE: unconventional, anti-establishment, decentralized, anti-
bureaucratic which 'challenged the predominant social paradigm of
Western industrial democracies' (Dalton, 1994: 19). As noted
above, this characterization is more applicable to FoE (US) rather
than FoE (UK).

Kitschelt (1993: 15) writes about how left and libertarian social
movements invoke 'an ancient element of democratic theory that
calls for an organization of collective decision-making referred to in
various ways as classical, populist, communitarian, grass-roots or
direct democracy against a democratic practice in contemporary
democracies labelled as realist, liberal, elite, republican or representative democracy'. He addresses the relationship between established democratic patterns and the direct democracy of social
movements.

Hager (1993: 53) also suggests that the role of citizen initiative
groups was not to represent the public in government but to participate directly in politics themselves: self-determination was the aim
of their activity. He claims that environmental groups, women's

groups, and peace movement organizations often avoid profession-alized, centralized, and bureaucratic organizations in favour of a fluid organizational framework (Hager, 1993: 101). Dalton (1994: 48) says that a stress on participation and self-direction is another trait identified with ecology groups (but in a footnote he makes the important concession that alternative societal goals do not neces-sarily translate into participatory norms within ecology groups). Dalton tries to maintain a distinction between conservationist and ecological groups in terms of participation:

ecologists ... advocate changes in the political structure of Western democ-racies, leading to a more open political system, a greater opportunity for citizen input, and a more consensual style of decision-making. In contrast, conservation groups often began as small, elitist associations. Even when these groups become mass organizations, they lack the emphasis on partic-ipatory politics often found in ecological groups.

Here, in our view, is the dangerous implication that the characteris-tics of small activist organizations are also found in large-scale group organizations. There is a conflating of the characteristics of a small number of ideologically self-conscious activist participants with the scale of those who are considered to be environmentalists on the strength of indicators no more profound than picking up other people's litter. Dalton gives the orientations of Robert Hunter, one of the founders of Greenpeace as: 'We need to move neither further to the Left nor further to the Right – rather, we must seriously begin to inquire into the rights of rabbits and turnips, the rights of soil and swamp, the rights of the atmosphere, and ulti-mately, the rights of the planet' (Hunter 1979; cited in Dalton, 1994: 49). Not only are these features not strongly associated with the supporters of larger groups, the fact is that the grassroots of large-scale groups are not *into* the 'rights of the soil or swamp'.

Generally, it is argued that ecology-type groups are less likely to be centralized and that weak opportunities for member input are linked with conservation-type/environmental-type bodies. If the generalization stands it may be that while overall it is true it does not hold for the major examples (certainly in neither Greenpeace nor FoE are there strong participatory opportunities at national level). One FoE organizer said in interview in 1995:

Members have to decide to back us or not. We make policy and if they don't like it they can join some other group. We have some members who are unhappy with our nuclear energy position but we are not going to change it just to be more popular.

Dalton's (1994: 245) perception of FoE is thus very different from ours. He says: 'At one extreme, one sees organizations in which members are simply passive supporters of the group, with no formal role in defining the organization's activities. FoE illustrates a contrasting pattern ... one that stresses participation and attempts to involve the members through local chapters and FoE-sponsored activities.' Our picture of FoE support is of 'couch' participation.

In fact, Dalton (1994: 106) concedes that in practice most environmental groups are not immune to the Weber-Michels imperative towards centralization: 'organizations need authoritative decision makers, and such authority nurtures oligarchy'. Shaiko's (1993: 90) article on Greenpeace USA begins by drawing attention to Greenpeace International's 1992 net operating budget of $25 million and a *membership* of 1.8 million in the United States. However, Bosso (1991: 167) argues that the use of the term *membership* is a misnomer – it is constitutionally and politically inaccurate. Contributors to Greenpeace are not formal members, but simply financial sponsors. Large organizations, such as Greenpeace and FoE, are best seen as having *supporters rather than members*. As Shaiko (1993: 93) points out, Greenpeace is like virtually all other US environmental groups which 'remain addicted to direct-mail as a mobilization and retention tool'. He cites a study by *Forbes* magazine that showed that Greenpeace USA raised $64 million in 1990 with 60 per cent coming from 40 million solicitations sent out by the direct mail firm Craver, Matthews, Smith and Co. If the social movement idea does not adequately capture the FoE type of organization nor does it much resemble the stereotype of the voluntary, active-member interest group. It appears closer to a protest business.

'Extreme' interpretation of pressure group: voluntary, democratically accountable, and individual-based

An interest group is popularly seen to have certain (stereotypical) defining characteristics:

1   It is organized *only* for a specific collective political end (such as the abolition of slavery).

2   Its goal is attainable; the group can be disbanded on realization of the goal.
3   It is a non-governmental body.
4   It does not seek itself to form a government, merely to influence public policy.
5   It has a formal voluntary membership of individuals.
6   The membership has some control over the leadership of the organization in connection with goals and means (internal democracy).
7   The membership funds the organization through subscriptions.
8   It is organized to give expression to shared attitudes or pursue shared interests (promotional/sectional).
9   As membership is seen to reflect careful matching of group and individual goals, then it can be expected that members join groups with some long-term attachment (see Jordan, Maloney, and McLaughlin, 1992: 14–16).

In fact, the organizations dealt with in this volume have sufficient common characteristics to suggest that there is an organizational form that is very different from a traditionally conceived interest group based on voluntary, policy-making members. As Hayes (1986: 135) observes, the traditional assumption was that groups involved face-to-face interaction, but in modern-style groups this simply does not happen (see Table 2.3). This is in line with what we term protest businesses (see Table 1.6).

Table 2.3 *Hayes' typology of interest groups*

Primary sources of financial support			
		Membership	*Outside sources*
Opportunity for face-to-face contact	High	Pure membership groups	Subsidized solidary groups
	Low	Mass groups	Pure staff groups

Source: Hayes, 1986: 137.

Pure membership groups are similar to the 'extreme' stereotype set out above. They are heavily reliant on members for funds and have an 'extensive network of local chapters that provide for realistic opportunities for rank-and-file intervention and influence on group decision-making' (Hayes, 1986: 138). These groups are seen as important 'mediating institutions in pluralistic societies'. *Mass groups/organizations*, on the other hand, are controlled by a small clique in a centralized location (e.g. London or Washington, D.C.) and are also heavily reliant on membership subscriptions for organizational survival. Hayes (1986: 139) cites Ralph Nader's Congress Watch as a mass group because members 'have little or no opportunity for interaction with other members ... (and the group is) dependent on direct mail solicitation for the bulk of it revenues. Congress Watch consists largely of a small, Washington based staff, with Nader making the key decisions himself.'

While recognizing a considerable overlap between our protest business and the mass group we prefer to retain our term for two main reasons. First, there are important links to business practice that are signalled by our term. Secondly, the term mass group echoes the concept of 'mass party'. This implies membership with a policy-making role. This is absolutely not the case with mail order groups.

The organizations we are examining in this book are closer to interest/pressure groups rather than elements in a (new) social movement, but above all are closer to *protest businesses*. Their hierarchical organization, lack of internal democracy, political strategies and tactics all undermine their categorization within the NSM rubric. This chapter has essentially attempted to trawl the social movement literature to measure the 'fit' between our examples and the (amorphous) social movement ideas. Dalton (1994: 3) notes that to supporters the green movement 'is seen as the vanguard of a new society' and 'environmentalism ... a fundamental threat to the established socioeconomic system'. Our basic conclusion is that there have to be serious reservations about citing the data we have about specific organizations in aid of discussions of an environmental social movement. In fact, we see FoE as an imperfect interest group not because it shares many of the putative features of SMOs but because it shares features of the very different protest business idea.

However both interest group and social movement literatures have come to share a reservation about the Olson argument which suggests that participation is the conclusion of a self-interested cost-benefit exercise. Fireman and Gamson (1979: 8) argued that those sociologists studying social movements should 'beware of economists bearing gifts' (cited in Morris and Mueller, 1992: 7). Also from the social movement perspective, Ferree (1992: 32) sees rational choice as a one-dimensional view of rationality that presents 'decontextualized individuals ... ambivalence, altruism and emotional experience are thereby made invisible and irrelevant'. This is also the attitude of non-rational choice political science. There are possibly two literatures that are addressing similar themes in similar ways. Others might conclude that this is in fact one literature and that the superficial distinctions add little to understanding.

Notes

1 Lowe and Goyder (1983: 3) contrast three different perspectives on environmental groups: the *organizational perspective; the social movement perspective*; and the *pressure group perspective*.

2 In this text we use the terms interest group/pressure group interchangeably: they are used to discuss the same phenomenon.

3 As Dalton *et al.* (1990: 4) point out it was a German sociologist Brand (1982) who 'coined the term "Neue Soziale Bewgungen" ... and the term "New Social Movements" (NSM) entered the English research vocabulary as an identifier for this new type of interest organization'.

4 What is the depth of their critique of the economic system? We suspect that it is not as fundamental as many commentators have argued – especially for the large-scale supporters of environmental organizations. It is probable/arguable that their support for this altruistic cause would evaporate were dramatic lifestyle changes or a substantial redistribution of wealth advocated.

5 Zald and McCarthy (1987: 119) point out that resource mobilization theory 'downplays the importance of formal membership criteria in favor of a more realistic notion of sympathizers, adherents, and constituents. Adherents and constituents may provide a variety of resources; yet they may not be formal members.'

6 Zald (1987: 321) points out another difference between 'new' and 'old' social movements: 'In Gamson and Tilly's terms', some of the early movements sought 'to gain access to the polity, to gain standing (e.g. voting rights, standing as legitimate members of the polity). But it is already the case for the movements based in the middle class that the issue is not polity membership, but policy influence and preferences.'

7 We are sceptical of the robustness of the distinguishing characteristics outlined in Table 2.1. For example, the archetypal 'old' social movement is the labour movement, but it could not be conceived as solely existing in the 'polity', it had roots in civil society. In addition, it was also involved in direct action via grassroots protest in pursuit of its objectives.

8 Kitschelt (1993: 528) sees social movements as arising when

aggrieved groups cannot work through established channels. Existing avenues such as parties and interest groups are closed and there is the emergence of unconventional, disruptive, and sometimes violent collective mobilization. This also is an interpretation of the social movement idea that locates the SM as a sort of influence strategy.

9   Dalton *et al.* (1990: 11) also argue that: 'The populist and participatory values of new social movements also stand in sharp contrast to the bureaucratized, hierarchical, and neo-corporatist tendencies existing in the most established European interest groups.'

10   Rucht (1990: 161) differentiates strategies from tactics. A strategy 'refers to a conscious, long-range, planned, and integrated conception of an actor's conflict behaviour based on the overall context (including third parties and potential allies), and with special emphasis on the inherent strengths and weaknesses of the major opponent'. On the other hand, tactics '… can be understood as a specific concept of conflict behaviour based on a situational assessment of available resources as well as of the benefits and costs of various forms of action, both for the actor and its opponent(s)' (Rucht, 1990: 174). Rucht (1990: 174) then cites Jenkins (1981: 135): 'Since tactics pertain to immediate situational contingencies, little of general value can be said about these decisions. This is not true, however, for strategy.'

11   Kitschelt (1993: 14) defines a pure public good as intangible in the sense that individuals cannot claim particular shares of the good in the same way that distributive collective goods can be pocketed by the recipients of pensions, higher wages, medical benefits, farm subsidies, etc..

12   Dalton (1994: 11) discusses NSMs in terms of *ideologically structured action*. The different identities adopted reflect and constrain the kind of political options adopted by the groups. Thus, Earth First! can be distinguished from the RSPB on the grounds that the latter 'accepts the dominant social order'. The RSPB is embroiled in the consultation process, while Earth First! relies largely on more unconventional political strategies.

3

How bumble bees fly – accounting for public interest participation

The political system is beset by a swarm of organizational bumble–bees that are busily flying about in spite of the fact that political scientists cannot explain how they manage it.

(Walker, 1991: 77)

Political science discussion of interest group membership in the last 30 years has been dominated by Olson's *The Logic of Collective Action* published in 1965. The literature has in large part been responding to his key proposition that: *'rational, self-interested individuals will not act to achieve their common or group interests'* (Olson, 1965; 1971 edn.: 2, original emphasis). Olson believed the limitations of the economics-derived approach to be remote and his rational choice perspective seemed to rule out the successful mobilization of public interest groups. Thus groups, particularly public interest groups, appear to operate in defiance of Olson's theory.

Olson's discussion about group membership rests on two principal types of incentives – selective and collective (Olson, 1971 edn: 51). The pursuit of collective goods will not, in his view, secure rational membership: potential members will free-ride. Rational individuals will decline to contribute to the costs of collective action unless there is coercion (e.g. a trade union 'closed shop'). He argued that: 'Only a *separate and "selective" incentive* will stimulate a rational individual in a latent group to act in a group-oriented way' (Olson, 1971 edn: 51, original emphasis).[2] Johnson (1995: 6) summarizes the Olson expectation as: 'A person will, make a contribution only when the personal cost is exceeded by the sum of selective incentives and the change in the personal valuation of the collective good that results from the person's contribution.'

Olson (1971 edn: 2) conceded that there was a logical possibility that groups comprised of either altruistic individuals or irrational individuals may sometimes act in their common or group interests. However, he went on: 'But, as later empirical parts of this study will attempt to show, this logical possibility is usually of no practical importance.'

Prior to Olson's contribution the most influential propositions about group development ignored the problem of free-riding. Truman's *The Governmental Process* advanced the 'proliferation' thesis – as social and economic differentiation occurred, more potential groups emerged – and the 'equilibrium' theory which suggested that economic or social change led to a disequilibrium in the set of organized groups; consequently new organizations emerged to re-establish the balance (Truman, 1951: 25, 31). In a prescient argument Truman (1951: 14) said that the 'Robinson Crusoe hypothesis that members are best conceived of as isolated units is inadequate psychology as well as unfashionable economics'. Nonetheless Olson rejected this broadly pluralist perspective, arguing that many interests would not find effective expression. His thesis did not simply examine the mechanics of group mobilization; it challenged important assumptions about the nature of the political system. Consequently, if Olson's thesis is flawed then access in the political system might be less restricted than he allowed. This has consequences for the critique of pluralism.

There is, then, a mobilization paradox: Olson's theory predicts the under-mobilization of public interest groups but this seems contradicted by their empirical proliferation. Contemporary interest groups have strikingly high memberships (e.g. the RSPB has nearly 1 million members). How do we account for the proliferation of interest groups of all kinds – and the particular expansion of public interest groups (for instance, on environmental, ethical, or consumer issues) which pursue collective goods? As Walker (1991: 61) argued citizen groups most sharply face the organizational dilemmas highlighted by Olson. Groups pursuing broad public goods have potential members that are geographically dispersed, the issues are marginal to the individuals involved, and selective material incentives are not likely to be available.

Our central proposition is that potential members do not engage in the intellectual gymnastics of the *economically* rational choice approach, costing their contribution against the likely personal return on investment. Such an exercise may well be intellectually impossible because the benefits are so difficult to quantify: and certainly the sheer effort of the calculation appears to be more than it is worth (see also Marsh, 1976). Or to put the argument more cautiously, the public interest group expansion and proliferation in the past decades suggests that sufficiently large numbers of sympathisers do not engage in such reasoning. Thus, the problem is how to amend the theory to relate to the practice.

This chapter highlights the limitations of the rational choice accounts of public interest group membership: it challenges the belief that the 'by-product' and 'imperfect information' theses on interest group membership are adequate to account for the scale of the phenomenon. Our data suggests that 'members' believe that campaigning is more important than Olson's model allows. The chapter also reviews non-Olsonian explanations of group membership and argues that non-material, *soft*, incentives (i.e. purposive, solidary, and expressive incentives) offer a more realistic perspective. We also set out how several different lines of explanation explain away the paradox of participation – indeed by specifying rational action in ways other than that set out by Olson the paradox disappears. Finally, we argue that there has been an *over-emphasis* in the literature on demand-side explanations. Groups successfully *market* (supply) themselves to their *predisposed public* (this is considered at length in chapter 6).

Understanding public interest group motivations: competing explanations

Olson (1971 edn: 64) saw membership on unselfish grounds as non-rational. In this chapter we identify several different explanations for group proliferation in general, and public interest group growth in particular. Cumulatively these explanations restore rationality to membership.

Olson's explanations

The by-product theory

Olson recognized that self-interested members did join organizations under certain circumstances: his explanation for this (abnormal) behaviour stressed the selective incentive. Olson's *by-product* notion was that: 'Only such an organization could make a joint offering or "tied sale" of a collective and a non-collective good that could stimulate a rational individual in a large group to bear part of the cost of obtaining a collective good' (Olson, 1971 edn: 134). This assumes that the provision of selective (material) incentives is central to membership decisions, and that this sort of membership incidentally provides the resources for lobbying for collective purposes. This interpretation is commonly advanced in connection with organizations such as trade unions which negotiate collective benefits such as general wage increases, but attract members through selective material incentives. For example, the British public sector trade union, Unison, offers its members 'free legal advice on any civil, matrimonial or criminal matter, (and) free financial advice' (*Independent on Sunday*, 5 September 1993).

Selective benefits tied to a collective appeal are also offered by some public interest groups. McFarland (1976: 4) notes that members of the Sierra Club can easily recover their (then) $25 dues through reduced – fare travel, hikes, and social get-togethers. (But he also notes that few of the members actually exploit these selective opportunities – throwing doubt on the idea that they are the important factors.)[3] Mitchell (1979: 122–9) pointed to the National Audubon Society's group insurance plan, and the various US environmental groups which offer a regular journal or magazine as a selective incentive. He does not, however, consider these to be Olson-type incentives because they are likely to be secondary rather than primary in motivating membership. In addition, these communications have a very important role for the group as a mechanism for membership retention. The cost of providing journals and magazines is seen by the group as a cost of recruitment/retention, rather than a selective incentive for the member. These communications are an important (marketing) weapon in the highly competitive membership battle between groups.

The marketing industry distinguishes between the *front-end premium* which is a minor inducement given to all the recipients of the mail shot (such as stationery with an appropriate wildlife theme) and the *back-end premium* (for example, a free wildlife video) which is the inducement only for those who contribute (Johnson, 1995: 17). The latter is the sort of material incentive that Olson stressed. Though some part of the proliferation of public interest

groups may be ascribed to the by-product theory and the effectiveness of these selective incentives, we doubt their importance.[4] A
selective incentive cannot be a major factor in causing support of a
public interest group though it can be a useful part of the mechanics of translating a general sympathy into active support. The
National Trust, for example, has an exceptionally large membership
(over 2 million in the early 1990s), a large proportion of whom
have joined for the selective benefit of free entry to the Trust's property. The fact that *most* of its members join *solely for* the selective
incentive, and that political campaigning is a low priority for its
members, raises a question mark over its status as a public interest
group. The more 'public' the group, the less powerful the selective
incentive.[5]

   For public interest groups the by-product explanation is unconvincing. Though material and selective incentives are found in this
area it seems unlikely that they are central to the joining decision.
In an important critique of Olson, Udehn (1993: 249) poses the
question that if selective goods are necessary, why bother with collective goods at all? He suggests that it is 'slightly absurd' that a
theory of interest groups sees the collective good issue as marginal
to selective incentives.

Olson's misguided members? imperfect information

Rothenberg (1992: 20) notes that potential group members do not
have perfect information (see also Moe, 1980a, 1980b). This interpretation can be expanded to rationalize apparently irrational
behaviour. Imperfect information might lead potential members to
participate – falsely believing that they are personally efficacious
and that their participation helps secure a public good (see Finkel *et
al.*, 1989; Muller and Opp; 1986). This thesis does not sit comfort-
ably with our evidence which shows a *bias* in participation towards
those with higher educational attainment: for example, 45.3 per
cent of new and 56 per cent of veteran FoE members have degrees,
as do 52.5 per cent of Amnesty members. Moreover, our data shows
that campaigning group membership is not a one-off error. Members tend to *repeat their 'mistakes'*: 65.8 per cent of FoE and 73.5
per cent of Amnesty members are members of other environmen-
tal/campaigning organizations.

   In reality, even if individual contributions have a limited impact on outcomes, groups may *persuade* potential members to the contrary. As Moe (1980b: 33) argues, group leaders need not simply leave the estimation (of personal efficacy) to chance, but can play an active role in *shaping* how individuals arrive at their decisions. Amnesty advertising deliberately confronts this issue:

Maybe you didn't think that anything you could do would make a difference. If Peter Benenson had thought that 32 years ago, Amnesty International would not exist today ... One person started that. One person who just could not sit back any longer and let things go on. One person is no different to you. If he could do that, think of the difference you could make by joining like-minded people. Imagine the help and support you could bring to others.

If individuals accept the 'guidance' from the group that membership does make a difference, they may be rational but misinformed. But why should the educated be particularly gullible? It seems more likely that these contributors are deliberately and consciously not behaving in an economically rational way. This leaves open the possibility that *other types of rationality* might underpin behaviour. Green and Shapiro (1994: 29) note that when Olson's predictions appeared to be falsified, rational choice theorists typically move to imperfect information models as the first line of defence – and if this fails the next step is to accounts that appeal to motives other than narrow self-interest.

Table 5.4 shows that FoE and Amnesty members consider that their support *has* an effect on FoE's/Amnesty's 'ability to protect the environment/protect human rights': 71.2 per cent of new, 74.7 per cent of veteran FoE members, and 69.6 per cent of Amnesty members believe that their support made a difference. Few respondents felt that their participation had no effect (7.3 per cent FoE veterans, 6.9 per cent of Amnesty members). These responses suggest that group members do not accept the free-riding premise; they believe that their support is an important element in the provision of the collective good. (Of course this does not prove that *active* supporters differ from non-mobilized supporters because of these beliefs. The non-mobilized might also share these views but we lack authoritative surveys of potential members.)

One possible reason for these beliefs is the success of what Dunleavy (1991: 74) terms 'size manipulation' strategies by groups:

which communicate partially contradictory messages about the group's size and viability to members ... interest groups stress that they are large, viable, nationally organized and worth joining. Yet simultaneously potential members are told that the group is small and local enough for their participation to make a difference; or at least their non-participation will be visible and noticeably impair group effectiveness.

Olson's imaginary individuals react to a large membership as an opportunity to free-ride, but the groups' instinct that leads them frequently to boast of a large membership would seem to have its own logic. Advertising a large membership reassures the potential member that it is sensible to be a member: 'Can 100,000 others all be wrong?'

The fact that some groups (especially public interest groups) experience a substantial drop-out of members at the end of the first year of membership, could be held partially to substantiate the 'imperfect information' thesis – as members learn more about the group, they lose interest (Rothenberg, 1992: 21). However, our data on FoE's lapsed members (Table 5.4) does not confirm this. Only one-quarter (26.1 per cent) of lapsed members think that their participation has no effect. Lapsing rarely results from deliberate rejection. An internal survey in 1993 by Amnesty of 4,000 lapsers found that only 4 per cent of respondents expressed 'disenchantment' as a reason for leaving.

Explanations denying the paradox of participation

Outflanking Olson: patronage

The first two sorts of explanation – selective incentives and imperfect information – are compatible with Olson, but arguably fail to match up in importance with the scale of actual membership. This chapter goes on to review other types of explanation.

Walker (1991: 75) argues that interest groups survive in ways not anticipated by Olson[6] (see also Berry, 1977, 1984; Salisbury, 1975). He found that almost 90 per cent of citizen groups received patronage assistance in their formative stages. Groups seek patronage from a financial foundation or other major financial contributor in addition to individual contributions. In the extreme form, this can obviate the need for members: Berry (1977: 28) showed that 30 per cent of public interest groups in the United States had no members. In

1993 approximately 60 per cent of the Prison Reform Trust's income (in the UK) came in the form of patronage from charitable trusts, foundations, and private companies. In most groups the role of external finance is to reduce the level of subscription needed from individuals. The invitation from the American Political Science Association (APSA) for members to rejoin in 1995 said: 'In addition, you get real value for your membership dollar. Because of other sources of income, including outside grants and interest on our endowment, membership dues pay for less than half the costs of APSA programs' (15 February, 1995).[7] Reducing the subscription in turn makes the propensity to join more likely – taking the decision below a threshold of rationality.

Walker (1991: 48) maintained that groups with large memberships which do not provide selective incentives or employ coercion attract mainly those with good education and ample incomes. Our data supports this: 44.2 per cent of FoE and 53.0 per cent of Amnesty members had household incomes over £20,000 (see Table 4.7). This suggests that for those with higher discretionary disposable income joining is less problematic and for them membership can be seen as a donation rather than an investment.

A high-income membership need not reflect greater sympathy for the causes in question among the better off: poorer potential members have more competition for resources and will not join because they lack 'free' resources (Barry, 1970: 34). This skewed membership is also an artefact of group advertising strategies. Groups aim their adverts at those with higher income, particularly using the quality press. One Greenpeace official told us that 'the tabloids are not a good recruiting ground'. The pattern of membership is generally reflective of the pattern of group marketing (see chapter 6).

Non-material (soft) incentives as an explanation of membership

Public interest group activity casts doubt on the material, self-interested decision as the explanation for membership. Clark and Wilson's (1961: 130) work anticipated Olson and put forward a general theory of incentives for 'organization membership' – i.e. in 'political interest groups, corporations, trades unions, universities, political parties, administrative agencies'. They distinguished three broad categories of participation incentive. First, *material incentives*, which are (basically) monetary rewards such as salaries. Secondly, *solidary incentives*, which are usually seen as stemming from the opportunities to attend meetings and to interact with others (charitable organizations (e.g. Rotary, Lions, etc.) might raise funds for good causes but the primary membership incentive is the solidary reward of being a member). Thirdly, *purposive incentives*, which are intangible benefits like solidary incentives, but which relate to the supra-personal purpose/success of the organization rather than the personal reward in acting. There is a fundamental difference between the solidary and the purposive. If one is attracted by solidary incentives one cannot free-ride the organization, but one could free-ride an organization whose purposive ends one supported, though it is psychologically improbable that one would wish to free-ride a cause of which one approved – if the cost was low.

Salisbury extended Clark and Wilson's typology by adding *expressive incentives*. He noted that not all so-called purposive incentives are the same. He suggested that in practice some of the attraction of Clark and Wilson's purposive goals may also be material – e.g. call for tax cuts or price. Salisbury (1969: 16) said that: 'Expressive actions are those where the action involved gives expression to the interests or values of a person or group rather than instrumentally pursuing interest or values ... benefits are derived from the expression itself.' This (again) is not 'free-rideable'. Hirschman (1982:88) labels this urge as 'striving' and claims that this is the sort of activity that gives our lives meaning. Irrationality in this light is not financially supporting the goal of Amnesty, but denying oneself the satisfaction of expressing support.

Organizations offer a menu of incentives to attract potential/predisposed members because they have learned that members join for a variety of reasons. (One group organizer we talked to used the phrase 'lots of little hooks'. In other words, different individuals join for different combinations of different incentives.) It may well be that the supporter of a particular cause group sees the gratification obtained by contributing as more important than any selective *material* incentive. There is a distinction between joining Amnesty as a deliberate act aimed at helping it secure its goals (purposive) and a contribution based on a(n) (expressive) wish to get satisfaction through contributing. This latter assumes that the act of contributing is a benefit in itself. There is a 'feelgood' factor that acts as a selective incentive. It is not 'free-rideable': as is the case with the solidary incentive the individual is rewarded selectively by this style of participation irrespective of the actual success of the group. These observations are not new. As early as 1981, Moe picked up Clark and Wilson's ideas and made the claim that an individual's joining decision can be shaped by incentives other than economic gain. The mystery is why these propositions are still controversial.

Parry, Moyser, and Day (1992: 9) identify four 'primary models' of participation – *instrumental, communitarian, educative* and *expressive* (we discount the educative type as that appears to be a statement about why individuals *should* participate). Instrumentalism involves participation '... intended to promote or defend the goals of the participants with the minimum of costs and the maximum of effect'. The goals may be altruistic of self-interested or a combination of both. The benefits of participation are seen to exceed the costs of such participation in a rational way. Within the instrumentalist perspective there are two variants that roughly match an important distinction in this chapter – *socio-psychological* and *economic*.

The socio-psychological approach emphasizes the centrality of individuals' 'civic' attitudes towards participation. Social background and life-chances affect an individual's predisposition to participate. Thus these 'civic' attitudes are associated with those who have higher socio-economic status (SES). Verba and Nie (1972: 12) say 'those with higher income, higher education, and higher status occupations', and who know more about the political process, tend to have higher levels of participation. In other words, it is rational for that part of the population to participate – even though the rationality might not be self-interestedly economically based (though as important stake-holders in society there might also be an economic dimension).

The 'economic' school of thought argues that individuals act in 'very strict instrumental terms and assess the values of public involvement in terms of the likelihood of achieving their objectives, compared with the time, energy, and frustration which could be anticipated' (Parry, Moyser, and Day, 1992: 10). Consequently, those with higher SES have more to gain by participating and more to lose by not, while the poor and disadvantaged in society simply see political participation as 'not worthwhile'.

The communitarian model sees participation as motivated by a strong concern for communities and assumes that individuals develop a sense of *interdependence* with other members of their community and 'seek to act to sustain their communal relationships' (see also the 'equity ethic' idea, below). Participation is a reflection (rational in its own terms) of investment in the community. Participation in modern societies is seen as 'relatively low' because of excessive centralization and the decline in the sense of community which has discouraged widescale participation (Parry, Moyser, and Day, 1992: 13).

Finally, there is the expressive school of thought that was introduced by Salisbury (above). Under this model people do not only participate to achieve an identifiable goal or out of a strong communal sense, but as a way of expressing their feelings or position on a particular matter. This assumes that the act of contributing is a benefit in itself. There is a 'feelgood' factor that acts as a selective incentive. This participation ranges from the largely symbolic and weakest sense of participation – e.g. putting an Amnesty or FoE sticker on your car – to the strongest sense 'the desire to be there' (Hardin, 1982: 108) – e.g. wishing to take part in major 'events such as the civil rights marches in the USA' (quoted in Parry, Moyser, and Day, 1992: 15).

The key to understanding the lack of coherence in the literature is that different authors have adopted different positions on what incentives are relevant in membership decisions. Olson's economic rationality focused almost exclusively on material incentives. The history of the literature is of a battle to re-insert non-material incentives into the calculation. The most obvious explanation for action which does not conform to Olson's economic model is that there are non-economic inducements operating – i.e. action is rational, but rationality is not the self-interested economic calculation set out by Olson. Moe (1980b: 188), for example, points out that an individual may derive a sense of satisfaction from contributing, when he or she sees this as an act of support for valued goals. Olson (1971 edn: 61) accepted that: 'In addition to monetary and social incentives [social status and social acceptance], there are also erotic incentives, psychological incentives, moral incentives, and so on.' He did not, however, pursue these for two main reasons: (i) they were simply not needed as his other incentives explained the cases; and (ii) they were non-quantifiable and consequently untestable. He said that:

'At no point in this study ... will any such moral force or incentive be used to explain any of the examples of group action that will be studied. There are three reasons for this. First, it is not possible to get empirical proof of the motivation behind any person's action; it is not possible to definitely say whether a given individual acted for moral reasons or for some other reasons in some particular case. A reliance on moral explanations could thus make the theory untestable. Second, no such explanation is needed; since there will be sufficient explanations on other grounds for all group action that will be considered. Third, most organized pressure groups are explicitly working for gains for themselves, not gains for other groups, and in such cases it is hardly plausible to ascribe group action to any moral code. Moral motives or incentives for group action have therefore been discussed, not to explain any given example of group action, but rather to show that their existence need not contradict the theory offered here, and could if anything tend to support it. (Olson, 1971 edn: 61–2)

Opp (1986: 88) refers to non-material economic factors as *soft incentives*:

Economists ... concede that there are other kinds of incentives. Olson mentions social and moral inducements; Tullock introduces ... the entertainment value of participation in revolutionary activities; Breton and Breton refer to the prestige and power which may be the rewards for participating in social movements, and Mitchell mentions ... self-esteem and sociability. These kinds of costs and benefits may be labelled 'soft incentives' because the utilities are not attached to material phenomena.

Udehn (1993: 248) argues that if Muller and Opp's 'doubtful' and 'sensational redefinition' of self-interest includes altruism, feelings of solidarity, and conformity to norms, then these factors cannot be ignored in the decision process, because they stretch the economic model.[8] Udehn (1993: 251) himself says of those who want to keep a narrow and parsimonious notion of rationality that: 'Like some religions, it may lead to other worldliness; to a withdrawal from the complexity of the mundane world, into a more simple world of pure abstraction. In this enchanted world of intellectual constructs, many social scientists feel at home, apparently satisfied with activities, such as model-building and story-telling.' Petracca (1991: 289) argues that 'rational choice omits far too much from the complex scheme of political life to be entirely reliable and useful as either explanatory or predictive theory'. *The 'pay-off' of membership may not be in material benefits.* The rational calculation is amended by incorporating a broader range of incentives. Below we note the work of Ferree (1992) and others who do not so much want to 'fine tune' this kind of calculation as to abandon it. Hirschman (1986) says that the persistent failure of simple models of rational choice to fit empirically cannot be disguised by relabelling everything as some form of self-interest or incentive (quoted in Ferree, 1992: 43). Our view is that though an expanded conception of rationality may well become descriptive rather than explanatory, it highlights worthwhile empirical data not available through the 'first principle' reasoning of Olson. As Udehn (1996: 256) shows, *mixed motivations* seem necessary for an understanding of collective action.

Table 3.1 *The three most popular reasons for joining FoE/Amnesty*

Reason	Weighted FoE sample (%)	Amnesty (%)
General concern for the environment/human rights	31.8	47.6
To support an environmental/human rights organization/to get involved/fight for human rights	13.7	13.5
To support Friends of the Earth's/Amnesty's aims, objectives and principles	11.6	22.1

Table 3.1 shows the three most popular categories of response to the open question: *Why did you join Amnesty International/Friends of the Earth?*[9] Other answers (less frequently) were similarly 'non-material' in nature. For example, FoE members gave among other reasons: 'financial support for a cause I believe in' (4.8 per cent); 'to be kept informed about environmental issues, (6.7 per cent); and 'to apply pressure to the government/big business' (3 per cent). In the United States McFarland (1984: 47) found similar responses for reasons for joining from Common Cause members. Twenty-eight per cent of his respondents said that they joined Common Cause because they 'wanted to have a say in government', while 23 per cent joined because they 'believe in its aims'. This pursuit of non-material ends is not a rejection of rationality *per se*, but a rejection of the narrow Olsonian economic version. Members contribute because the value ascribed to the return of selective and collective

material *and* non-material incentives exceeds the membership cost.

This expansion in the factors considered as rational exposes tensions in the definition of rationality. Salisbury (1969), for example, sees all membership (short of mental illness?) as rational.[10] Rational membership is explained by factors other than material self-interest. The variables that Salisbury takes into account in explaining membership (e.g. 'good government', and 'peace') are not easily translated into an economic calculus. Moreover, the factors considered relevant in the decision – even if they were calculable – are really calculable only by the potential member. No outside observer can contradict the assessment about membership and decide if it was 'really' rational or not.

It is not always clear how to operationalize the Olson theory – and how behaviour is to be 'scored' in his terms. For example, one tactic used by some groups is to 'adopt an X' gambit. Johnson (1995: 12) notes that the International Wildlife Coalition, for example, administers the Whale Adoption Project. He describes how a person who adopts a whale receives a personalized whale adoption certificate, a photo of the whale, a four-colour migration map, a decal, and a subscription to *Whalewatch*. Johnson suggests that such programmes fit in with the Olsonian perspective in that they suggest that the impact of the individual's contribution to securing the collective good is significant (i.e. the subscription is – or at least appears to be – 'pivotal'). Of course the pivotality is illusory, or more precisely is an illusion constructed by the group. As Johnson notes, the same whale is adopted by several, perhaps thousands of others though this reality is not too obvious in the direct mail literature. He concedes that this sort of membership is largely congruent with the Olson approach. Members seem as keen on a package of selective 'goodies' as much as a collective good. In his survey of groups 73 per cent of the group leaders said that members expected something *tangible* in return for contributions. The implication is that these members are selective and material in their expectations. In fact, the newsletters and other material can be part of an assurance that the collective good is being effectively pursued rather than a selective material incentive. The individual wants to know that she or he personally is doing good, not that she or he wants personal goods.

One interpretation of what is going on is that the donor is misguided in over-estimating his or her role. Another is to assume that their eyes are open and they are in fact contributing knowingly to a collective good without much sense of pivotality. A further possibility is that they are giving because they enjoy giving: an expressive reward. This makes sense whether or not whales are in fact saved by the contribution.

Sabatier (1992: 105) rejects extending the analysis to cover non-economic incentives because it leads to a tautology: members join because there are reasons for joining. Expanding self-interest to include the personal satisfaction from altruistic acts, he argues, renders the entire enterprise non-falsifiable; consequently it should be abandoned. However, the baby should not be thrown out with the bath water. There may be utility in allowing soft incentives into the calculation. It may be tautological to say that the outcome is a variety of factors but this is more useful than attempting to explain outcomes by one inadequate (economic) dimension. As Barry[11] (1970: 34) highlights there is an argument that: 'in the absence of economic or social "selective incentives" ... the reason for joining has to be altruistic'. Our data confirms that members claim not to be (simply) interested in their immediate personal welfare: 65.1 per cent of FoE members and 64.4 per cent of Amnesty members stated that '*join(ing) like-minded people in fighting for the environment/ human rights*' was a very important/important reason in their decision to join. Just because both material and non-material factors are involved is not an argument against the relevance of the latter, but rather it means that the decision process is no longer testable in economically rational terms. As Cook (1984: 424) argues: 'Although the benefits are most often non economic and intangible in nature, the members clearly indicate they have expected to gain something from their public interest group in exchange for the costs of affiliation.'[12]

Our 1993 survey collected data that refers to NFU Countryside. It is in essence a fund-raising venture, set up in 1992, by the National Farmers' Union to raise income by selling services and providing information for non-farmers living in the countryside. It secured 5,000 members in the first six months. The organization motto is *Serving the People Who Care*. Members are seen as caring for the countryside, but the nature of concern might be very different from that of members of FoE. NFU Countryside claim that membership 'has attracted an amazing variety of people; from avid gardeners, devoted horse lovers, rare breeds enthusiasts, through to dedicated conservationists and organic growers' (NFU Countryside advertising, December 1992). Discounts are very much in the front of the 'shop window' in the advertising for recruits. It is pointed out that 'NFU discounts on goods and services are now available to NFU Countryside members.] These range from 20 per cent off Canon faxes and word processors, to a low price on an NFU first aid kit, to 10 per cent off office equipment, discounts on TV sales and rentals, car hire, and a Travel Club.

The NFU Countryside data demonstrates that some groups are much nearer to the Olson model than others. While FoE and Amnesty can be seen as offering a selective incentive to their members in the form of publications (e.g. *Earth Matters*) which are only available to members, we – like Berry (1977) – believe that these publications do not represent much of a selective incentive.[13] However, in addition the NFU Countryside provides information and advice notes on issues such as husbandry, new businesses, and countryside law are provided via free-post order forms. The comparison of NFU Countryside with Amnesty/FoE underlines the different sort of attractions that groups can offer. Many members of the NFU Countryside are in membership for the material incentives offered (most especially, NFU Mutual Insurance) but there are also overlapping interests with other more mainstream public interest bodies. For example, the RSPB has contributed to the *Countryside Journal*. The journal also publicizes news items such as the availability of CPRE/Transport 2000/World Wide Fund for Nature guide to campaigning against road proposals. Members get fact sheets from the Farming and Wildlife Advisory Group on matters such as 'Caring for the Environment', 'Tree Planting', and from the Rare Breeds Survival Trust and similar organizations.

The appeal of collective goods rather than material self-interest

Contrary to Olson's by-product assumption, our evidence in chapter 5 and in line with other work (see Godwin and Mitchell, 1982; Cigler and Hansen, 1983; and Schlozman *et al.*, 1995b) suggests that members cede more importance to the satisfaction obtained from supporting campaigning than Olson suggests (see Table 5.5). The provision of resources to allow the pursuit of collective goods is not a by-product, but is central to the decision to join. We 'tested' the by-product proposition by asking members and lapsers of FoE if they would rejoin/renew their FoE membership if: (a) direct member services were reduced?; (b) core campaigning was reduced?

The results were that 77.6 per cent of new and 87.8 per cent of veteran FoE members said that they would remain in membership if FoE *reduced direct member services* in favour of core campaigning, but only 32.9 per cent of new and 31.3 per cent of veteran members said that they would remain in membership if *core campaigning was reduced*. (Only 5.9 per cent of lapsed members would be persuaded to rejoin if core campaigning was reduced (see Table 5.5)). While these responses need not be an accurate guide to actual behaviour, they are nevertheless suggestive of the importance members (and lapsed members) attach to FoE's political campaigning activities. Our data supports Hansen's (1985: 94) findings that: 'For some people political benefits are sufficient of themselves. Others join both for services and for policy but either alone is insufficient. For still others, political benefits are the crucial quality difference' (see also Moe, 1980a: 607). Olson (1979: 149) argues that such types of answer may simply be reflecting a socially acceptable response, but in the light of the lack of any other obvious consideration, this is unduly sceptical.

The equity ethic as a membership explanation

The previous section of this chapter moved to discussing the value group members placed on collective rather than selective incentives. In effect, acceptance of this sort of motivation by-passes the paradox of participation that emerges from an economically rational choice perspective. 'Hard' incentive theory does not 'work', but needs to be supported by a range of other factors.

One of Olson's (1965) supporters, Hardin (1979), concedes that in talking to a brighter than average part of the population (an undergraduate class at the University of Maryland) the majority thought that the proposition that it was 'irrational' to support a labour union was crazy. Students, and others, found nothing odd with the idea that: 'One puts one's money where one's mouth is' and contributes towards ends one supports. To them Olson's position is very much special pleading. In the spirit of seeing Olson as counter to a common-sense approach. Green and Shapiro (1994: 79) give the example of a committed Christian fundamentalist walking through a park in which two demonstrations were going on simultaneously – a pro-choice to the left and a pro-life to the right. As they point out, the Olson analyst would argue that since the individual calculates that participation in either movement is unlikely to influence desired policy outcomes she might look who was providing the best snack. As their (deliberately absurd) example implies, because she has no incentive to sacrifice the opportunity cost of a good snack she might join the pro choice-rally were the catering good enough!

Walker (1991: 47) summarized Marwell and Ames' (1979) study which suggested that there has been a rise in 'equity norms'. They discovered that in experimental situations there were normative ideas of 'fairness' that stopped free-riding (see also Sabatier, 1992: 125). There is evidence which shows that the public is not simply interested in lowest cost purchasing. The 'fair trading' concept offers customers the opportunity to buy products that help Third World producers (e.g. goods in Oxfam catalogues). This view of motivation recognizes an ethical impulse which drives potential members to contribute. McFarland (1976: 6) refers to the 'civic-balance beliefs' of well-educated citizens who may participate in organizations to balance the power of special interests. This is a sort of intervention in support of civic duty.[14]

Udehn (1993: 243) cites work that treats the pursuit of collective action not as a one-off decision but as an 'iterated Prisoner's Dilemma supergame'. As is well known, in a one-shot Prisoner's Dilemma the best outcome for an individual is to look after their immediate self-interest and not to co-operate – though if all players do so they will all be worse off than if they had co-operated. However, if the event recurs then conventions can be established that are mutually beneficial. Axelrod (1984) and Taylor (1976) (both cited in Udehn, 1993) show that co-operative strategies can pay -off. This point perhaps suggests why groups tend to advertise their success in recruiting members – though in Olson's terms that is an invitation to free-ride. The groups are in effect saying that X thousand others are contributing and the onus is for other supporters to fulfil an implicit contract.

An apparently strange empirical detail (contrary to Olsonian orthodoxy) picked up by Johnson's (1995) survey of US environmental organizations is that the generally trivial 'front-end premium' sent to all non-members contacted is a more powerful incentive to response than the 'back-end premium' offered selectively to those who reply. Omitting the latter can cut response rates by up to a third but omitting the former can reduce it by two-thirds. Johnson says that the reason might be that the front-end factor demonstrates the sort of usefulness of membership, but at least as persuasive an argument is that the gift creates a sense of obligation that touches the equity ethic impulse.

Collective bads and minimax regret strategy

While Olson is sceptical about the pursuit of collective goods as an incentive for public interest membership, some commentators have distinguished between the motivating power of the pursuit of collective goods and the avoidance of collective bads. Mitchell (1979: 115) argues that: 'The primary cost of not contributing is the continued existence of environmental bads, and more important, the possible increase in environmental bads in the future.' Thus individuals who value environmental goods but fail to contribute potentially incur a cost if the group fails to obtain these goods due to insufficient funds.

King and Walker (1992) also see membership as more likely in a threatening environment and cite the increased mobilization of business groups from the 1960s – in reaction to consumer success – as evidence of this. It is clear that group advertising tends to emphasise 'bads': it has been found that negative images stimulate more applications. Olson[15] might well argue that one can free-ride the protection from collective bads just as one might free-ride the expected provision of collective goods. Mitchell (1979), however, maintains that a stronger motivation exists to avoid bads than exists to seek improvements. He suggests that contributors to environmental groups might be following the same rationality that Ferejohn and Fiorina suggested underpinned voting – the so-called *minimax regret strategy*. This runs along the lines of: 'My God, what if I didn't contribute and because of the lack of my contribution the group was unable to prevent the emasculation of the Clean Air Act, get the Alaska Bill passed in Congress, etc.' (Godwin and Mitchell, 1982: 163). Thus membership can be driven by guilt avoidance rather than a belief that there will be a positive outcome hinging on the individual contribution. Instead of maximizing expected utility, individuals choose to minimize maximum regret.

Commitment theory

Sabatier and McLaughlin (1990) argue that participation stems from 'beliefs about good policy' (cited in Sabatier, 1992: 109). The expected benefits *may* be material but they are commonly a mixture of material and ideological or purposive benefits. This perspective assumes that although material self-interest can create an incentive to join a group, only those who perceive large material benefits, or who have buttressed self-interest with ideological incentives, will be sufficiently committed to play an active role in the organization. Accordingly, members come from a particularly committed subset of the potential group membership. Sabatier (1992: 110) suggests that commitment theory:

agrees with the by-product theory in hypothesizing that only a small percentage of the beneficiaries of a group's political activities will be members, but it does so for radically different reasons ... most potential members will lack the material or ideological commitment to take the time and expense to join; most people are simply not very interested in, or informed about, policy issues.

The central conclusion of Sabatier's work is that Olson's assumption, that potential members (at least of purposive groups) who fail to join are 'free-riders', is too simple. Instead, he suggests that those who fail to join are less committed than those who do so. This may not be surprising but nor is it unimportant.

Udehn (1993: 254) notes that people do co-operate, even when rational egotism suggests free-riding, 'But not all of them do so to the same extent.' He picks up Elster's idea that some people operate as 'everyday Kantians'. They are prepared to co-operate unconditionally because it is their duty to do so. But, Udehn (1993: 254) says: 'If there are enough Kantians, the utilitarians might find it worthwhile to join, thereby creating the conditions for the norm of fairness to come into play. Therefore there might not be a universal rationality but people might act rationally in terms of their different belief systems.'

Different rationalities

The previous sections described how Olson's basic version of rationality has been expanded to incorporate non-economic variables. These explanations above are part of an expanded rational choice model and are likely to be criticized by the purist rational choice adherents who prefer the parsimony of Olson, and by fundamental critics of any rationality (e.g. Green and Shapiro, 1994) who see the expanded approach as degenerating into description. Several other explanations of public interest group membership can be grouped. They do not follow the Olson material/self-interest assumption, nor do they seek to extend the Olson approach by building in other considerations. Instead they incorporate the decision in another paradigm: this may imply another rationality but it is not explicit. Thus giving blood to the medical services for free in Britain is not economically rational: there is a cost of time with no selective material reward. However, while this makes no sense in terms of a *logic of consequentiality*, it is perfectly reasonable in terms a *logic of appropriateness* (March and Olsen, 1989: 23).

Rationality as conditional expected utility

One reaction to the paradox of participation is to argue that it is the formulation of rationality that is wrong: when corrected the paradox disappears. Olson's form of rationality form is of the broad expected *utility* type. This is a type or rationality, not *the* type. An alternative formulation is *conditional expected utility* (Jeffrey, 1983). Here the potential participants assume that their own inclinations indicate (as opposed to cause) the likely behaviour of others:

A Downsian does not see his own behavior as a source of information. Since it is his own behavior, after all, what could it tell him when he already took into account all relevant beliefs and preferences when deciding to produce it? A conditional expected utility maximizer, in contrast, sees her behavior on a par with other sources. It can be as informative as the behavior of other individuals. (Grafstein, 1994: 8)

Grafstein (1994: 5) reconstructs the calculus (in the context of voting) as follows: 'If I vote (join) then decision-makers with my preferences are more likely to vote (join), in which case the conditional expected utility of voting (joining) may be larger than its costs. Hence voting (joining) can be rational for those who maximize conditional expected utility.' Grafstein stresses that individuals do not believe that their participation will cause others to do so, they believe that if they feel themselves under pressure to participate others might act similarly under these conditions. Sen (1967) drew attention to the notion of 'assurance' that suggests that co-operation can be a sensible strategy under certain conditions, while Elster (1979) has written of a 'conditional preference for co-operation' (both cited in Udehn: 1993).

Margolis (1982: 103) sets out a 'fair share' model of rationality. Once this is accepted as a 'rational' priority then there is no irrationality in supporting public interest groups: 'In the case of a large charity raising many millions, Smith personally will not be able to see any difference his donation makes. But his motivation ... does not turn on any such perceptible return to Smith personally but on a judgement of social values.' Margolis thus argues that free-riding is not a rational activity if we allow for this refined sense of rationality. He observes that while the economically rational approach predicts the free-rider problem, this in practice is not as serious an issue as suggested because a different rationality is used. He says: '... no society we know could function if its members actually behaved as the conventional model implies they will'. Udehn (1996: 257) notes that one reason for Margolis advancing his alternative to standard public choice was the 'inability of economic man to solve the problem of collective action'.

New social movements

As set out in the previous chapter, for those writing from an NSM perspective the expectation of appropriate individual actions are very different from those in Olson's account. Dalton (1994: 8) describes the alternative orientation:

Although RM (resource mobilization)/rational-choice theory has advanced social movement research, there has been a growing awareness of the limits of this approach ... One impetus for this re-evaluation was a concern in enshrining rational action, the RM approach removed politics and values from the studies of mass movements ... a second impetus for the re-evaluation was the rise of a new set of contemporary citizen movements ... whose nature appear inconsistent with RM theory.

The NSM orientation sees no paradox in participation: participation is seen as natural as it was in the pre-Olson political science literature. As quoted earlier, Morris and Mueller (1992: 7) recall the words of Fireman and Gamson, that 'sociologists should beware of economists bearing gifts'. They say that the rational actor model has been a Trojan horse for social movement theory, bringing with it a radical individualism that presupposes a 'pseudo-universal human actor without either a personal history or a gender, race, or class position within a social history'. Melucci (1989: 31) rejects the political exchange approaches of Pizzorno and others, arguing that the motivation to participate cannot be considered as an exclusively individual phenomenon; it is, he argues, constructed and developed through interaction. Ferree (1992: 33) draws attention to moral commitments as being 'meta-rational' in the sense of Weber's 'value rationality' as opposed to what is termed 'merely instrumental rationality'. Ferree (1992: 41) cites work which rejects the notion of a single pseudo-universal rationality. This, it is claimed, turns out to be the values and perspectives of white, middle-class men in Western capitalist systems. Etzioni is quoted as making the point that such moral commitments are often explicitly based on the denial of pleasure in the name of the principle(s) involved. This sort of approach does not so much answer the problem of the Olson paradox, as deny it.

Individual modernity

Another approach that seems to address the fact of mobilization of those in society with educational and other advantages is advanced in the 'Beliefs in Government' project (Fuchs and Klingemann, 1995). Fuchs and Klingemann claim that by the mid-1970s there was a scholarly debate about the crisis of representative democracy which led to a sense that there had been societal changes which the political institutions had not accommodated. They note the work of Inglehart which identified economic growth, technological development, rising levels of education, and the expansion of mass communication as important. They quote Inkeles (1983: 12ff.; Fuchs and Klingemann, 1995: 12): 'An informed participant citizen; he has a marked sensed of personal efficacy, he is highly independent and autonomous in his relations to traditional sources of influence … he is ready for new experiences and ideas.'

If such a notion holds then the fact that the group participants are educated and confident in their efficacy is to be expected rather than treated as an anomaly in the Olson manner. In Table 5.9 we show that FoE and Amnesty members are indeed strikingly more 'post-material' in their values than the members of NFU Countryside.

Consumerism

Membership of environmental and campaigning groups may be interpreted as a *social signal rather than a calculation*. It may be an expression of self through the conspicuous consumption of membership via stickers and badges which advertise lifestyle orientation. Subscribing to these organizations is a statement, a piece of cultural identification (see Featherstone, 1991: 14). Bennie and Rüdig (1993) showed that the environmental attitudes and behaviour of the young were effectively 'fashion'-oriented commitments. Rüdig *et al.* (1993: 56) argued that the high turnover problems suffered by the Green Party in the 1988–93 period suggested that: 'some young people joined the Green Party as a political "fashion statement", and they left once the Greens had gone out of fashion'.[16] Parents of teenagers driven to purchasing designer label clothes know that the operant paradigm is not economic rationality. Political scientists may be reading too much into membership, certainly of large 'mail order' groups, if they see this as an act of *political* participation and evidence of an active citizenry.

This factor puts a gloss on the idea of the selective material incentive. If the material incentive is something that can be secured in the marketplace it is unlikely to be effective unless it is genuinely cheaper (e.g. insurance deals). But if the good has a symbol or motif that is exclusive to the group then a cachet might pertain that gives the item a profitable consumer value. Johnson (1995: 24) notes that Olson assumed that the collective good and the selective incentive would be separate but in fact they can be intimately linked. Thus the best selective incentive allows the supporter to feel linked with the collective goals of the organization. Johnson (1995: 24) says groups may have their best chance of competing with ordinary retail outlets if they link their selective incentives to the collective cause: 'by choosing to focus their collective-action effort and marketing effort on a particular creature, the group is able to convert a mundane item into something for which people are willing to pay. A group's identification with a collective good creates demand for certain kinds of selective incentives it may offer.'

Conclusions

The rational choice approach has been criticized as stressing methodological individualism and ignoring the influence of social structures (see Petracca, 1991: 293). This book maintains that the explanation of membership has to shift its focus: (i) from that of cognition by the individual to understand the decision in relation to the recruiting activities of the group (see chapter 6); and (ii) to accepting that a broad range of incentives – not just material and selective – influence joining decisions. It underlines the importance of the role the group plays in informing the potential member of its existence, in persuading the potential member that there is a problem to be addressed, that the group can remedy the problem, and that membership will enhance the group effort.

The *supply-side* perspective helps account for instability in public interest group membership noted above. It is the marginality of membership that makes the stimulus and the initiative of the group so vital: it acts to convert a broad but low intensity sympathy into membership. The (repeated) joining opportunities and the group's marketing strategies are more important in accounting for large numbers in membership than imagining that these decisions are the aggregate of thousands of economic cost/benefit analyses. Groups have a role in initiating membership – reducing the effort of joining, minimizing costs through patronage, and persuading potential members of the importance of the issue and the utility of their role. Group size or income is not simply a matter of the uncoordinated aggregation of individual decisions. Public interest groups have adopted the tactics of merchandizing organizations to stimulate the level of membership decisions (see chapter 6 for a full exposition of the marketing thesis).

Some would argue that to understand public interest group membership it is a mistake to use Olson as the datum. His explanation of group mobilisation resting on selective benefits has, some would claim, little utility in the cases of bodies such as FoE and Amnesty. One could decide to restrict the application of Olson's assumptions to the economic groups at the core of this thinking. However, Mitchell (1979: 88) notes that Olson considered that his perspective could 'be applied whenever there are rational individuals interested in a common goal'. Olson (1971 edn: 60–1) argued that 'people are sometimes also motivated by a desire to win prestige, respect, friendship, and other social and psychological objectives ... *social status and social acceptance are individual, non-collective goods*'. The acclaim accorded to *The Logic Of Collective Action* rested on interpreting it as a general purpose model. Restricting the argument to economic, self-interest groups cuts the work's importance – it is then no longer the missile aimed at the heart of pluralism.

The simplification that says that Olson is contradicted by the existence of public interest groups is, like all simplifications, inadequate. His free-rider proposition remains elegant. However, in relation to public interest groups of the FoE type there are many other factors in the joining decision. One can try to reconcile such decisions with the Olson approach by imputing an economic value to non-material rewards, but this is to depart from the simple, parsimonious explanation (see Marsh, 1994). It is difficult to resist the conclusion that public interest participation is often inspired by non-material incentives and influenced by a variety of social pressures and networks. Our data suggests that the public are prepared to invest in support of collective goals. There are non-material incentives which operate selectively. If the group has few material incentives to offer but prospers nonetheless, then members are receiving non-material satisfactions from membership.

Before Olson's contribution, the accepted perspective on membership was:

Olson's membership test

Join if:
Cost of subscription   < Value to potential member of
                              material benefits not obtainable
                              without membership

The revisionist literature implies a different calculation:

Expanded rational membership test

At any one point in time
Cost of subscription, and
time, and other costs   < Value to potential member of
                             selective material benefits

                             and/or

> Psychological benefits
> (self-esteem, status etc.)
>
> and/or
>
> Evaluation of role of contribution
> in securing collective goods
>
> and/or
>
> Evaluation of role of contribution in
> securing avoidance of collective bads
>
> (Matters such as the evaluation of collective bads or the pivotal-
> ity of a contribution towards the securing of the collective good
> are shaped by the group.)

Moreover, some argue that the whole economically rational choice approach is seriously one-dimensional and that to explain member-ship in terms of calculation is adopting a false image of the process: joining is the expression of belief. Despite our own acceptance of expanded rationality that includes a search for expressive satisfac-tion and other soft incentives, there is something intuitively wrong in 'improving' the Olson 'test' by making it more and more elabo-rate. This imputes to the potential members a far more 'laboured' (and elaborate) decision process than is perhaps appropriate. When we suggest that the decision has to be envisaged with a complex set of factors as shown above we do not thereby suggest the individu-als make a complex choice. They operate under rules of thumb that are evolved for low cost decisions.

Johnson notes (1995: 1) that a reputable poll in 1992 found that 67 per cent of the US public agreed with the statement that: 'Pro-tecting the environment is so important that requirements and stan-dards cannot be too high, and continuing improvements must be made regardless of costs.'[17] This allows Johnson to calculate that at least 120 million Americans are pro-environmental – yet, he points out, at the very most 15 million are in membership of relevant groups. Olson (1979: 149) argued that the study by Tillock and Morrison of Zero Population Growth (a group that existed with no selective incentives) supported rather than undermined his thesis because, he argued, this organization had only 12,000 members while those in favour of its ends was virtually everyone. Hardin (1979) similarly argued that the level of membership is significantly lower than the professed level of support for public interest causes. Hardin (1982: 106, quoted in Johnson, 1995: 2) suggests that the proportion of those actually mobilized into membership is so low that it does not represent much of a challenge to the strength of the Olson prediction, but it is a 'relatively small residual' of collective action that can be accounted for by moral or psychological motives such as altruism, guilt, or a sense of injustice at having to suffer collective bads in order to profit others.'

However, as Green and Shapiro (1994: 81) argue, to not-be-in-membership is very different from free-riding. This raises an under-exposed conflation in the notion of free-riding. Hardin and Olson seem to want to claim all sympathetic non-members as free-riders, but is one free-riding, if one has not heard of the group or if one is unconvinced about the effectiveness of the group? Indeed, it was Olson who highlighted the fact that one might sensibly not participate if one did not see the value in adding a tiny increment to the group resources. Is a rational non-joiner the same as a selfish free-rider? In trying to get a sense of the scale of free-riding it would not be sufficient to establish the percentage of the members among the sympathetic. One would need to establish how many see merit in group role and are nonetheless not joining.

Green and Shapiro say that even a very low rate of collective action may nonetheless involve enormously large numbers of participants. In our view membership rests not on the by-product proposition of Olson but nor is it non-rational as implied by Hardin. Whereas Hardin concedes that for some people certain soft incentives operate to a greater extent than that considered by Olson (or at least those who attempted to explicate his work) our position is that this special pleading to reconcile behaviour with the theory is not necessarily the solution. In fact potential members *can* choose to approach the issue in ways other than through economically rational analysis: for them it is not paradoxical participation because facts other than selective material incentives operate. And for others the notion of self-interested calculation is not present as they adopt other rationalities and work through other belief systems.

The economically rational understanding of political participation is still defended. Whitely and Seyd (1996: 143) claim that the rational choice theoretical paradigm has been so successful because it provides a powerful predictive tool for delineating how individuals behave in choice situations. They say that given the 'real successes' of rational choice it would be foolish to abandon the central notion of rationality in trying to understand political decision-making. However, their own paper in the edited collection concludes that while the rational model has important explanatory power, it provides an incomplete account of participation. They say that an alternative model which incorporates variables incompatible with a rational actor model does much better. At some point the stretching of the Olsonian model has to snap. *When does an extended Olsonian approach require a non-Olsonian conclusion?*

Notes

1 This chapter is based on an article by the same title in *Political Studies* 44(4) 1996.

2 Olson (1965, 1971) also noted that there are special conditions relating to small groups with relatively large members which allow them to mobilize successfully despite the free-rider problem. In these circumstances the large potential members may find it worthwhile to bear the burden of funding the activity.

3 The Sierra Club (in the United States) appears to provide an interesting example of an organization whose membership was previously attracted by selective incentives but which changed to one whose membership is attracted primarily by public goods. Mundo (1992: 174) cites the Club's Executive Director, Michael McCloskey as saying that: 'in the mid-sixties, three-quarters of the members joined for outings and one-quarter for conservation. By the early seventies, the ratios were reversed.' Mundo (1992: 174–5) added that: 'While Sierra Club members still enjoyed the information supplied by the club on outings, the impression of club officials and available survey data suggest that people joined to support the cause of conservation and environmentalism, and not so much to receive information and guidance on wilderness outings.'

4 Johnson (1995:17) found a respondent willing to give an empirically based 'feel' on these points. His source said that omitting the premiums entirely led to a 'marked decrease in response'. The group expected a response rate of around 1.5 per cent on a campaign, but the membership director said that without the selective incentives (back-end premium) this could fall to between 1.06 and 1.25 per cent.

5 The WWF offers a credit card with no annual fee paid by the user: the group gets a commission each time it is used. But one is unlikely to use

the card simply to save the fee: it is a means of expressing support for the goals of the organization. The slogan is 'Fund WWF projects at no extra cost to you.'

6   The patronage idea is compatible with, although not developed by, Olson.

7   Some groups finance their operations almost totally from subscriptions or personal donations, e.g. Greenpeace had a policy of not accepting funds from government or industry.

8   Udehn (1993: 252) treats the work by Elster (1989) much less critically but his 'mixed motivations' scope seems as broad as that of Opp.

9   The question asked was: *Why did you join Amnesty International/Friends of the Earth? (Were there any particularly important reasons or specific events which encouraged you to join?)* Note Johnson's argument (1995: 3) that membership studies that attempt to explain membership by reference to data from members commit a serious error in research design in that they can reveal traits common to members but not factors which distinguish members from non-members.

10   There is a major cleavage within the rational choice school between those who believe that all action is rational (by definition) and those who accept that certain actions can be deemed as non-rational. We follow Salisbury (1969: 15) who points out that: 'It should be understood at the outset that we do not attempt to assess "real" or "true" benefits. Rather we assume that people pursue those experiences and things which they value, for whatever reasons, and *in this sense only* may be regarded as rational. We assume that people mainly do or seek, subject to periodic evaluation or correction, whatever brings them a positive balance of benefits over costs.'

11   Barry (1970: 33) says that: 'Obviously, the constant danger of "economic" theories is that they can come to "explain" everything by re-describing it ... Thus if an organization maintains itself, we say "it must have provided selective incentives".'

12   Broadly speaking the debate about membership rewards has been conducted in terms of two categories, the public good and the selective incentives. In fact there are many semi-public goods that are not selective in the sense that they will be accessible only to group members but they are sought only (or at least wanted with intensity) by a proportion of the public. Thus control of smoking or the availability of abortion facilities are not selective but they are far from being public in the sense that everyone would pursue them or wish to avail themselves of them. We also need to distinguish between the goal as being the provision of a good and the provision of a principle. Many who campaign for abortion legislation are not campaigning for a personal matter but for their interpretation of a particular type of society. Therefore those who benefit are not those who have abortions but those who want to live in a society in which such opportunities

exist. That is to say that those who back abortion get far more from the legislation (even if they do not personally have abortions) than the unmotivated public. Indeed many in the public would fundamentally oppose such legislation. Campaigning for abortion is not campaigning for a straightforward public good but secures personal satisfaction for the campaigners and is opposed by other members of the public. The fruitful discussion of these sorts of activities in terms of only two types of incentive seems improbable.

13   Berry (1977) found that 76 per cent of the public interest groups he examined provided members with a publication, but few – 10 per cent – offered services in addition to this.

14   Schlozman *et al.* (1995b: 18) have distinguished between selective social gratifications and selective civic gratifications as well as selective material benefits and collective outcomes. Though they accept criticisms of individuals' retrospective reconstructions of motivations, their data looks robust and the conclusions are clear and well expressed: 'In short the data show that selective material benefits are quite infrequent and seem inadequate to explain the volume of political activity. We must look elsewhere. An obvious place is the other selective, but less tangible, benefits, of a social and civic sort ... Civic commitment on the part of ordinary citizens seems sincere, not a rhetorical cloak to cover some more calculated and self-interest motive.'

15   Johnson (1995: 8) is another who is unimpressed by the notion that threats are a distinctively different class of incentives. He notes that what may be happening is that publicity over the threat brings the issue to the attention of a group of people who would have donated in the past had they known about the group. Though this argument is in sympathy with the stress we ourselves place on publicity in chapter 6, we are sceptical. The overwhelming belief of group organizers in interviews (and as expressed through their advertisements) is that bad news is a good recruiting agent.

16   The Green Party made a net gain of more than 10,000 members between 1988 and 1990, and a net loss of more than 13,000 between 1990 and 1993 (this net loss represents loss of half its membership) (Rüdig *et al.*, 1993: 56).

17   Survey also cited by Bosso. See chapter 1.

4

Who participates in 'mail order' groups?

Groups with large memberships that do not provide selective material incentives or employ coercion to hold their members attract mainly those with good educations and ample incomes.

(Walker, 1991: 48)

This chapter reports data not on those who express concern about issues but on that proportion of the *predisposed publics* that are *in 'membership'*[1] of specific organizations: the characteristics of the sympathetic and the mobilized may not be the same. The organizations on which we focus have been termed 'mail order' groups (Mundo, 1992:18) in that the relationship between the individual and the organization is essentially financial. There is little face-to-face activity and, in general, members respond to press or direct mail solicitations.

This chapter, and chapter 5, compare our British data on Amnesty, FoE, and NFU Countryside with: data collected by Amnesty and FoE themselves; membership surveys from North America (Godwin, 1988; McFarland, 1984; Rothenberg, 1992); and political party membership in the United Kingdom (Rüdig *et al.*, 1991; Seyd and Whiteley, 1992; Whiteley *et al.*, 1994). Dalton's work (1994) – *The Green Rainbow: Environmental Groups in Western Europe* – which uses European-wide data is also used as a comparison. Comparison with party membership is important for the discussion of the decline of party thesis (as set out in chapter 7), and the claim that organizations like Amnesty and FoE have been the main beneficiaries of this process. Are there any implications for direct or participatory democracy which flow from these points (again discussed in chapter 7)? What are the implications for participatory democracy if the socio-demographic data shows a very narrow mobilization of the public in so-called public interest groups? Is the nature of the membership experience participatory in any very real sense?

Our earlier caution about distinguishing between mobilized members or supporters and those simply disposed to be sympathetic to public interest group goals is particularly important in connection with Dalton's contribution. Citing data based on surveys of mass opinion and data on actual groups may be confusing two different phenomena. There may be no close link between mass group memberships and much more specific, and perhaps active populations, such as the women's movement, peace groups, and the self-help movement (Dalton, 1994: 4–5). It *may be* true that those who are active in such social movements 'are calling for a new agenda for contemporary societies and demanding that governments open the political process to more diverse and citizen-oriented interests' (Dalton, 1994: 5); but members of large-scale interest groups, far less the weakly sympathetic, non-mobilized public, need not share these priorities. As Dalton acknowledges (1994: 72): 'we doubt whether most Europeans would endorse the more vibrant shades of green future proposed by the radical environmentalists'. In our view this distinction is often lost, and the 'big numbers' of the mass campaigning groups are often assumed to be evidence of the appeal of, and support for, a radical agenda.

In fact, participation in mail order type groups is characteristically slight in the effort implied. 'Members' are perhaps more accurately seen as financial supporters than policy-making participants. Though Amnesty has formal(/legal) member-based policy-making procedures, in fact few attend the AGM. However, a 1989 internal Amnesty survey showed significant levels of local participation: 21 per cent of members were active in local groups. Out of a total membership of over 100,000 some 9,500 wrote letters on behalf of Amnesty more than 12 times per year. Sixty per cent of local group members were not members of Amnesty International (British Section). Our 1993 data recorded that 15.7 per cent of national group members belonged to local groups. Cohen (1995: 173) in his study of Amnesty in Britain has noted that:

'Support' might imply even more minimal action than sending an annual donation. There are increasing opportunities for what might be called 'channelled acknowledgement'. Variations include ethical investments ('Profits with a conscience'); schemes where part of your regular credit card payments goes to a charity or buying politically correct products that you would anyway enjoy (like Ben and Jerry's ice-cream or The Body Shop's 'environmentally friendly' offers).

Basic demographics

Gender

Our data found a significant and largely unforeseen gender imbalance in membership (see Table 4.1). Women are disproportionately 'active' in public interest groups. The group findings are not, however, out of line with other sources about attitudes. Jelen, Thomas, and Wilcox (1994: 172, 180) have pointed out that research in recent years has shown that women 'take more liberal positions on various matters of public policy'. This gender gap is apparently largest for issues such as the environment. Rüdig et al. (1991: 16) have pointed out that in public opinion polls and the British Social Attitudes Survey, women are usually found to be more concerned about the environment, nuclear energy, and the threat of nuclear war. [2]

Table 4.1 *Gender differences in membership of FoE/Amnesty*

	Weighted FoE sample (%)	Amnesty (%)	NFU Countryside (%)	Resident UK population (1990)[a] (%)
Male	40.4	42.3	68.5	48.8
Female	59.3	56.4	30.6	51.2

Note:
[a] CSO, *Annual Abstract of Statistics 1992* (London: HMSO).

The data on gender imbalance appears to be robust – confirmed as it is by data collected by the respective groups. For example, Amnesty's own (1989) survey found a 47:53 male:female split and a 1992 FoE internal survey found that for a variety of membership categories women comprised the majority (see Table 4.2).

Table 4.2 *Gender differences for various categories of FoE members*

	MSO[a] (%)	Members (%)	Donors (%)	Lapsers (%)
Male	34	44	37	29
Female	66	56	62	71

Source: FoE survey, 1992.
Note:
[a] Monthly standing order.

Our figures for FoE members and lapsers highlight higher female mobilization. This data contrasts with that on party membership. Seyd and Whiteley (1992: 28, 98–9) found that 60.8 per cent of the members of the Labour Party were male, and that women members were less likely to be active than men. Thus, 19 per cent of women in the Labour Party claimed to be 'very active'; the corresponding figure for men was 28 per cent. Table 4.1 also shows that the NFU Countryside provides a striking contrast to Amnesty and FoE. This suggests the possibility that public interest groups – which do not provide material selective incentives – tend to attract more women than men, while the material incentives offered by the NFU Countryside are more attractive to men.

The phenomenon might relate to the fact that women have distinct policy preferences. Or the decision by women to act through groups may reflect a sense of relative disadvantage in other political arenas. Alternatively, it might simply be part of the self-reinforcing process in which groups recruit on the basis of their existing membership profiles (see chapter 6). In fact, though gender is a feature of membership there seem few consequences of this difference: female members are not distinctively different – except that their membership may be more unstable (the non-rejoining of young single women in their first year of membership may be as high as 60 per cent). Questions such as 'level of commitment to the organization' produce very similar responses for males and females. For example, among Amnesty members 66 per cent of men are very/fairly committed and the corresponding figure for women is 69 per cent.

The reasons male and female respondents gave for joining

Amnesty were also broadly similar. Thirty-four per cent of males and 42 per cent of females 'liked Amnesty's non-party political approach'; 21 per cent of males and 28 per cent of females believed they were 'joining like minded people'; and 22 per cent of males and 21 per cent of females joined because of 'Amnesty's responsible campaigning style'. The gender imbalance may be more interesting than politically significant. There was no very sharp difference on matters such as membership of local groups (plus 0.8 per cent male); taking part in urgent letter schemes (plus 2 per cent female); taking part in letter writing campaigns (plus 5 per cent female). Amnesty women are slightly more likely to be members of other environmental organizations and are consistently, if only slightly, more positive above the effectiveness of other campaigning/environmental organizations.

On the whole then, similarities rather than differences emerge from looking at the issue of gender. For example, 43 per cent of Amnesty male supporters feel closest to the Labour Party – and 40 per cent of women do likewise. Women are more likely to be Green Party 'leaners' but the level is distinctively low (9 per cent). Taking first and postgraduate degrees together there is broad balance among members, and 77 per cent of women perceive themselves to be middle class (when forced to choose between working-class and middle-class positions) as do 73 per cent of men.

Education

Ingram, Colnic, and Mann (1995: 123) note that the people who join environmental groups differ most strongly from the public as a whole in the level of higher education received. Our data also illustrates that membership is skewed towards those with higher educational attainment. Twenty-seven per cent and 37 per cent of new and veteran FoE members respectively have first (not 'first class') degrees, as does 26 per cent of Amnesty's membership. Even more surprising is the number of respondents with a postgraduate qualification. NFU Countryside members do not resemble Amnesty and FoE members with only 17 per cent having a degree qualification. Nevertheless, they are still relatively well educated when compared to the general population (see Table 4.3).

American surveys yield results comparable with ours. For example, McFarland (1984: 49) found that 43 per cent of Common

Cause members had completed a 'graduate or professional degree program'. As Rothenberg (1992: 31) points out, Common Cause members are an incredibly well-schooled bunch: 'Even more striking is that 55 per cent of group members – ten times more than the average citizenry – have received postgraduate education.' The possession of a degree qualification (or equivalent) is probably the most important individual resource of those who tend to participate in political activities. As Parry *et al.* (1992: 69) argue: 'degree-holders are not only an educational elite, but they are also a participatory elite'.

Table 4.3 *Highest educational qualification gained*

	Weighted FoE sample (%)	Amnesty (%)	NFU Countryside (%)	General population (%)
Degree	35.3	26.0	17.4	8[a]
Postgraduate	18.9	26.5	5.5	–

Note:
[a] Highest qualification level attained of persons aged 16–69 not in full-time education in Great Britain in 1991. (Bridgwood and Savage, 1991).

In our data there are a relatively high number of members with degrees, and 19 per cent and 27 per cent of FoE and Amnesty members respectively have a postgraduate degree. This sort of educational imbalance extends to lapsed members (29 per cent had degrees). The only difference is that *fewer* lapsers (9 per cent) had a postgraduate qualification; the corresponding figure for new and veteran members is 18 per cent and 19 per cent respectively. Nevertheless, lapsers still tend to be highly educated.

In terms of education Labour Party members more closely resemble Amnesty and FoE, while Conservative Party members are more like the NFU Countryside respondents. Twenty-eight per cent of Labour Party members (Seyd and Whiteley 1992: 238) and 12 per cent of Conservative Party members (Whiteley *et al.* 1994: 264) had a university or polytechnic degree or diploma. Those with higher educational attainment seem to have a greater sense of political efficacy. In addition to this, Inglehart (1995: 65) argues that they are also more likely to possess post-material values (see below).

Occupation

Table 4.4 shows that the overwhelming majority of FoE and Amnesty members were in professional, managerial, or senior administrative occupations. Just under half (48 per cent) of FoE members and just over half (52 per cent) of Amnesty members are 'professionals'. FoE's 1992 survey found a similar occupational profile to ours with 9 per cent of members in administrative/clerical occupations and 46 per cent in professional positions. In the United States membership of public interest groups reveals similar types of participants: for example Mundo (1992: 178) found that 70 per cent of Sierra Club members were in 'professional' occupations. Thus, other survey work in the United Kingdom and the United States confirms that this sort of bias to the professions for such groups is not remarkable.

Table 4.4 *Selected occupational categories for FoE, Amnesty and NFU Countryside members*

Occupation	Weighted FoE sample (%)	NFU Countryside (%)	Amnesty (%)
Clerical worker (clerk, secretary, telephone operator)	7.7	6.4	7.5
Professional or technical occupation (doctor, school teacher, engineer, social worker, accountant)	48.4	19.6	52.2
Manager or senior administrator (company director, manager, executive officer, local authority officer)	11.0	24.7	15.2

When looking at veterans and lapsers it appears that those most likely to remain in membership tend to be in professional occupations; however, a significant number of lapsers are also professionals: any difference is in the degree of professional dominance.

Occupationally NFU Countryside members are not distinct. They also tend to be professional/managerial. Only 8 per cent and 2 per cent of the NFU Countryside members described themselves respectively as farmer/farm manager or farm worker. This is not as surprising as it may initially appear. The NFU Countryside aims to recruit non-farmers living in the countryside; the farmers are catered for in the 'parent' organization (the NFU) (see chapter 3).

Class

Many FoE, Amnesty, and NFU Countryside members were reluctant to identify themselves in class terms. (When pressed to choose between working or middle class, FoE members and lapsers, Amnesty and NFU Countryside members overwhelmingly identified themselves as middle class: see Table 4.5.) Most of those who perceived themselves as belonging to a social class (unprompted) saw themselves as middle class (52 per cent of FoE, 51 per cent of Amnesty, and 44 per cent of NFU Countryside). This contrasts sharply with the findings of the *British Social Attitudes: 7th Report* which found that only 29 per cent of respondents considered themselves to be upper middle/middle class, and 65 per cent perceived themselves to be upper working/working class.

Table 4.5 *'If you had no choice but to place yourself in one of two classes, which one would it be?' (prompted)*

	Weighted FoE sample (%)	Amnesty (%)	NFU Countryside (%) (%)
Working Class	20.9	22.9	22.8
Middle Class	73.7	71.3	69.4

Seventy-three per cent of Labour Party members and and 62 per cent of Conservative Party members considered themselves to be members of a social class. However, 50 per cent of Labour Party members stated that they were members of the working class, with 20 per cent middle class, while 47 per cent of Conservative supporters considered themselves to middle class and 11 per cent working class (Seyd and Whiteley, 1992: 236; Whiteley *et al.*, 1994: 261). Labour Party members appear closer to the general population in terms of self-perceived class than members of the groups we surveyed.

Dalton (1994: 69) examined *Eurobarometer* survey data on the predisposition to join nature protection or ecology groups and argues that 'despite the theoretical link between the rise of the new middle class and the emergence of NSMs, class differences in environmental support are weak'. This appears to contradict our findings, but the figures are not strictly comparable: his data refers to *predisposed* members and ours relates to *actual* members. There may be some bias in mobilization that brings the middle class into prominence.

However Worcester's (1994: 10) data found a strong class factor on the environment along the lines we report. The average number of 11 'Green Activism' dimensions practised by the public was 3.49 but in the average AB (upper-middle class, professional, or managerial household) the figure was 4.87: 'ABs are more than three times as likely to do five or more of the named activities, the definition of the Environmental Activist, than are DEs.'

Looking at the Amnesty membership in class terms does not provide us with a very clear 'story'. The working class were more likely to join local groups (21:16 per cent), but there was a lead for the middle class in terms of 'taking part in letter writing campaigns' (49:37 per cent). Over 90 per cent of those in both social class categories intended to renew their subscriptions and both had a high degree of political efficacy: 64 per cent of the working class, and 71 per cent of the middle class believed that their participation had a 'significant or noticeable effect' on the group's ability to protect human rights. The middle class are also more likely to be members of other organizations (78:56 per cent); the working-class members were more likely to vote Labour (65 per cent).

Age

Bennie and Rüdig (1993: 14) found that in the United Kingdom 23 per cent of those under 21 make financial contributions to environmental associations. This was nearly twice as much as youngsters in Europe as a whole and more than double the (probably more affluent older) generations. The data in Table 4.6 tends to confirm their argument. It shows that Amnesty and NFU Countryside members tend to be older than those of FoE. Amnesty's age profile is closest to the general population. For example, 51 per cent of Amnesty and 62 per cent of NFU Countryside members are 45 or over as opposed to only 31 per cent of FoE members. There is little difference between lapsers and members of FoE in terms of age (28 per cent of lapsers are over 45). FoE is predominantly an organization of the young, indeed 62 per cent of its new members are under 35 (and 82 per cent are under 44); the Amnesty figure is 25 per cent and the NFU Countryside is 14 per cent. The Amnesty internal survey in 1989 found 54 per cent of its members were under 35 (and 72 per cent were under 45). An FoE internal survey also confirmed our findings: 59 per cent of members and 68 per cent of MSOs (those paying by monthly standing order) were under 45. FoE's 1992 survey found that Lapsers tended to be 'young': 28 per cent were under 26 and a further 30 per cent were between 26 and 35. Bennie and Rüdig (1993: 19) concluded that:

in the United Kingdom, there is evidence that many young people took up 'the environment' as a fashionable issue of the late 1980s but that their commitment to environmental practices and policies remains rather shallow. Buying 'green' products was clearly 'chic' and was practised by a large majority of British youth; a commitment to less glamorous activities such as making less noise and using less energy, however, is comparatively lacking.

This could be a strong component in explaining FoE lapsers' failure to renew their membership.

Income

Forty-four per cent of our FoE sample and 53 per cent of our Amnesty sample had household incomes over £20,000 (see Table 4.7). There is little difference in income distribution patterns between new, veteran and lapsed members of FoE. Our data is comparable with some of that collected in the United States. For example, McFarland (1984: 48–9) found that: 'The median Common Cause member had a family income of about $37,000 at time when the national average family income was about $20,000 per year.' Mundo (1992: 178) found that 66 per cent of Sierra Club members had household incomes over $50,000 (but 23 per cent of the United

States population had household incomes over $50,000). This perhaps suggests that those with greater discretionary disposable income see the threshold where they are reluctant to contribute as higher. As Loomis and Cigler (1986: 18) argue, this more affluent population are more interested in association with organizations reflecting post-material values, and post-material values are greatest amongst those with higher incomes and higher educational attainment i.e. the type of individuals who are members of FoE and Amnesty.

Table 4.6 *Age profiles of Amnesty, FoE and NFU Countryside members*

Age group	New (%)	Veteran (%)	Lapsed (%)	Weighted FoE sample (%)	Amnesty (%)	NFU Country- side (%)	General population[a] (per 000)
+65	4.1	12.0	6.3	10.8	17.1	8.6	20.7
45-64	11.8	21.5	21.8	19.9	34.0	52.9	23.6
35-44	20.1	23.1	26.9	22.6	21.3	22.8	13.7
25-34	40.3	35.6	25.6	36.3	23.5	12.8	16.2
15-24	21.4	7.1	16.8	9.4	1.7	1.4	12.8

Source: CSO (1992), *Annual Abstract of Statistics*, 1992: 12.
Note:
[a] Mid-year estimates of the age distribution of the resident UK population.

Table 4.7 *Total household income distribution in the membership of FoE, Amnesty and NFU Countryside*

	Weighted FoE sample (%)	Amnesty (%)	NFU Countryside (%)
Under £10,000	18.1	14.9	9.6
£10,001–£20,000	33.3	28.2	26.9
Over £20,000	44.2	53.0	56.2

In terms of income, FoE, Amnesty, and NFU Countryside members are not clearly distinguishable from members of the Conservative Party, but they are different from Labour Party members (see Table 4.8).

Table 4.8 *Total household income distribution in categories of FoE 1992 members, and Conservative and Labour Party members*

	Members (%)	Conservative (%)	Labour (%)
Under £10,000	16	26	38
£10,001–£20,000	31	34	33
Over £20,000	47	42	29

Source: FoE Survey, 1992; Seyd and Whiteley, 1992: 253; Whiteley *et al.*, 1994: 45.

Parry *et al.* (1992: 77) found that: 'The wealthiest quarter, and especially those in the top 5 per cent participate at relatively high levels, whereas the poorest quarter are well below average ... The "haves" seem to participate more, and the "have-nots" less' (see also Rosenstone and Hansen, 1993: 13).

Party orientation

Though Amnesty and FoE supporters are party members to a greater extent than the general public, they are not to any great degree Green Party members. Nor, indeed, are they Green voters. Only one Amnesty respondent claimed to be a member of the Green Party, while 2 per cent FoE members claimed Green Party membership. Rüdig *et al.* (1991: 33) found that 52 per cent of Green Party members were also members of FoE; 51 per cent were members of Greenpeace; and 24 per cent were members of Amnesty.

FoE and Amnesty members are closest to parties of the left, and indeed, tend to vote for such parties: 29 per cent of new members and 28 per cent of veteran members (of FoE) and 42 per cent of Amnesty members voted for the Labour Party in the 1992 General Election (see Table 4.9); while 27 per cent, 31 per cent, and 31 per cent of new, veteran and Amnesty members respectively voted Liberal Democrat in 1992. Seyd and Whiteley (1992: 92) show that 16 per cent of Labour Party members were also 'members' of Greenpeace, 8 per cent were 'members' of FoE, and 7 per cent were 'members' of Amnesty. Evans *et al.* (1994: 12) have argued that Labour and Liberal voters tend to share similar 'attitudes towards libertarian/authoritarian new agenda issues ... which suggests those two parties are likely to be competing with each other for the progressive vote'. Public interest groups such as Amnesty and FoE are also in a competitive market in their attempts to attract these individuals into membership.

Table 4.9 *'Which party did you vote for in the 1992 general election?'*

Party	Weighted FoE sample (%)	Amnesty (%)	NFU Countryside (%)	General election[a] (10 April 1992) (%)
Conservative	11.1	9.9	62.6	41.8
Green	10.0	5.2	3.2	0.6
Labour	28.0	42.3	2.7	35.2
Liberal Democrats	30.3	31.2	14.2	17.0
Nationalists	0.2	1.7	0.5	2.3
Other	4.0	2.2	0	3.2

Note:
[a] CSO (1993), *Social Trends 23* (London: HMSO).

Table 4.10 *FoE membership. How important was it that you were 'joining like-minded people in fighting for the environment'?*

Party	Very important (%)	Important (%)	Not very important (%)	Played no role whatsoever (%)
Conservative	15.8	42.1	15.8	26.3
Green	15.0	40.0	40.0	5.0
Labour	29.6	42.3	23.9	4.2
Liberal Democrat	25.5	41.5	27.7	5.3

This finding about group participation being associated with clear party preference is confirmed in work by Aarts. Aarts (1995: 249) concludes that: 'There is no support for the hypothesis that (potential) activists in new social movements are ideologically different from other people. As might be expected, they are more likely to prefer left-wing or green parties ... That is not to say that supporters of new social movements do not have a distinct political orientation. What we do find, is that their party preferences fit well within the traditional framework.'

Table 4.10 shows that significant proportions of party supporters who were members of FoE said that joining a group of like-minded people was an important factor in their decision to join.

Overlapping membership

A key proposition of this book is that the distinctive pattern of group membership need not reflect greater sympathy among those with higher income. The poorer potential members have more competition for resources and will not join because they lack 'free' resources. The group's marketing is aimed at those with higher income. As Rosenstone and Hansen (1993: 36) highlight: 'Few people participate spontaneously in politics. Participation, instead, results when groups, political parties, and activists persuade citizens to take part.' *The pattern of membership is generally reflective of the pattern of group marketing.* In short, well-organized groups get the members that they seek. As the Market Research Manager at the RSPB informed us:

as a general rule we have been targeting people with a similar profile to our existing member base ... More recently though, we have observed encouraging results from warm-name conversion. In particular, this has been the case with people who responded to an advertisement for our countryside campaign [emphasis added].

For both FoE and Amnesty the main recruitment effort comes from direct marketing approaches to those on purchased lists (of names/addresses) with what are considered appropriate demographic qualities, or exchanged with other organizations that have comparable appeal. Table 4.11 shows that 17 per cent of Amnesty members are members of FoE, while 13 per cent of FoE members are Amnesty members. We can also see that 32 per cent of FoE members and 34 per cent of Amnesty members are members of Greenpeace. FoE's own survey found that 39 per cent of members, 38 per cent of those on MSOs and 37 per cent of lapsed members were also members of Greenpeace; and that 17 per cent of members, 18 per cent of MSOs and 18 per cent of lapsers were also members of Amnesty. The fact that our FoE and Amnesty samples are so similar is not surprising: both organizations contact not just the same sort of people but actually the same individuals (in fact, membership lists are exchanged).

Table 4.11 *Overlapping memberships*[3]

	Amnesty (%)	Oxfam (%)	FoE (%)	Greenpeace (%)	National Trust (%)	RSPB (%)	RSPCA (%)
Weighted FoE sample	12.8	5.7	n.a.	31.9	4.6	7.8	2.2
Amnesty International	n.a.	12.2	16.9	33.7	4.7	9.1	2.2
NFU Countryside	0	0	2.3	3.2	8.2	5.0	0.5

The advantage of exchanged or purchased names is that the groups know that those on the lists have already responded to a mail order appeal: it is assumed that such individuals are more likely to 'repeat buy'. The group will market test a sample of potential new lists to discover the likely profitability of an appeal to that list. The concept of 'lifetime value' of different types of member obtained from a new list is used to identify the 'best' lists in the sense that some will appear to give a better 'take up', or a 'take up' that is particularly strong among the sorts of recruits that will stay in membership to allow the recouping of the recruitment costs or give more than the minimum subscription. [4]

Thirty-nine per cent of NFU Countryside members said that they were members of other 'conservation or environmental' organizations. This contrasts with the FoE and Amnesty findings where the overlap was even higher: 56 per cent of new, 68 per cent of veteran and 50 per cent of lapsed members of FoE said they were members of other 'campaigning organizations'; and 74 per cent of Amnesty members claimed to be members of other environmental and campaigning organizations. A 1989 survey by Amnesty found that 29 per cent of their membership were also members of Greenpeace, 13 per cent of FoE, and 7 per cent of WWF. While the NFU Countryside figure is high in relation to participation in general, it is significantly lower than that for FoE and Amnesty. There is also a difference in the type of organization which NFU Countryside members are likely to join, or indeed feel sympathetic towards. While NFU Countryside members tend not to be members of organizations such as Amnesty, FoE, and Greenpeace, they are generally (and possibly surprisingly) sympathetic to such organizations. Table 4.12 shows that NFU Countryside members have a relatively high regard for these organizations (sympathy towards Greenpeace is probably the most surprising finding). But they are much stronger supporters of the conservationist wing of the environmental movement. For example, 85 per cent and 78 per cent of NFU Countryside respectively felt very sympathetic/sympathetic towards the RSPB and the RSPCA.

Table 4.12 *NFU Countryside members: 'How sympathetic do you feel towards each of the following organizations?'*

Organization	Very sympathetic (%)	Sympathetic (%)	Not very sympathetic (%)	Not at all sympathetic (%)
Amnesty International	11.0	38.8	32.4	11.0
Friends of the Earth	14.2	38.4	22.8	18.3
Greenpeace	14.2	30.6	25.6	23.7
Green Party	3.2	17.4	33.8	38.8
National Trust	31.1	45.7	11.9	7.3
Royal Society for the Protection of Birds	38.4	46.1	8.2	3.7
Royal Society for the Prevention of Cruelty to Animals	46.6	31.1	10.5	8.2

Member profiles

The typical 'member' of FoE/Amnesty is likely to be a well-educated middle-class female under 45 in a professional/managerial occupation from a relatively affluent household, who is a member of other campaigning organizations (most notably Greenpeace) and votes for a centre-left party (Labour or Liberal Democrat – although Amnesty members are more likely to vote Labour than Liberal

Democrat). Lapsed FoE members share a remarkably similar socio-demographic profile to members. There are, however, some minor differences. For example, lapsers are overwhelmingly female (over 70 per cent) and are less likely to be members of other campaigning organizations. In contrast, the typical NFU Countryside member is likely to be a relatively well-educated (although much so less than the FoE/Amnesty member), middle-class male over 45 in a professional/managerial occupation from a relatively affluent household, who tends not to join campaigning organizations (although feels sympathetic towards these organizations) and who might well vote Conservative.

Notes

1 While those who send annual contributions to Amnesty are members as conventionally understood, the status of those sending money to FoE is different. Their financial contributions give them no policy-making role in the organization.

2 More recently, Schlozman *et al.* (1995a: 288) found that in the United States there was a striking similarity between both the participatory modes and the participatory rewards of men and women: 'Across a wide variety of participatory acts, we found few significant disparities in men's and women's levels of participation ... women do not participate disproportionately in grassroots, organizational, local, ad hoc activities ... (and) men have no monopoly on material benefits ... the contours of women's and men's participatory agendas bear striking similarity.'

3 It is also worth noting that 7 per cent of FoE members are members of both Amnesty and Greenpeace; and that 8 per cent of Amnesty members are members of both FoE and Greenpeace.

4 Shaiko (1993: 94) notes that (for the United States): 'It is known that multiple membership is the norm among individuals who join environmental organizations.' (He also points out that it is not clear that Greenpeace donors understand or acknowledge the significant ideological differences between the Greenpeace leadership and those of other organization to which they belong.)

5

The dynamics of group membership

Even if the member of a large group were to neglect his own interests entirely, he still would not rationally contribute toward the provision of any collective or public good, since his own contribution would not be perceptible. A farmer who placed the interests of other farmers above his own would not necessarily restrict his production to raise farm prices, since he would know that his sacrifice would not bring a noticeable benefit to anyone. Such a rational farmer, however unselfish, would not make such a futile and pointless sacrifice ... Selfless behavior that has no perceptible effect is sometimes not even considered praiseworthy.

(Olson, 1971 edn: 64)

This chapter explores reasons that might account for the pattern of membership reported in the previous chapter. In addition to examining the differences between FoE/Amnesty and NFU Countryside members, it also explores the factors which explain why FoE supporters failed to renew their 'membership'.

When?

Table 5.1 illustrates that the vast majority of members of both FoE and Amnesty joined their respective organizations in the ten years prior to the survey: 85 per cent and 78 per cent of FoE/Amnesty members have joined since 1985. FoE membership may have been stimulated by: the growth of environmental concern in the United Kingdom as measured by opinion poll data; the 'flash' success of the Green Party at the 1989 European election (in fact, approximately 50 per cent of FoE members and lapsers joined between 1989 and 1991 inclusive); the increasing perception of organizations such as

FoE and Amnesty as 'successful'; and a higher media profile and marketing strategy capitalizing on these developments.

Table 5.1 *'When did you first join FoE/Amnesty International?'*

Year	FoE new (%)	FoE veteran (%)	FoE lapsed (%)	Weighted FoE sample (%)	Amnesty (%)
Before 1975	0.0	1.4	1.3	1.2	0.3
1975–79	0.0	6.0	1.3	5.0	1.5
1980–84	0.0	10.0	3.8	8.5	10.8
1985–89	0.0	43.8	37.5	36.7	65.0
1990–93	100	38.8	34.9	48.6	13.0
Missing	0.0	0.0	21.4	0.0	9.7

Why?

Table 5.2 shows that 81 per cent of FoE members were 'very/fairly committed' to the organization, while 56 per cent of lapsers were as 'committed'. It is not surprising that members are more committed than non-members, but what is surprising is the level of commitment asserted by lapsers after they had left the organization. Table 5.3 shows that over 90 per cent of Amnesty and FoE members are 'definitely/probably' likely to renew their membership. This is a surprisingly high figure given that these organizations are likely to experience an annual drop-out rate up to 30–40 per cent.

Table 5.2 *Degree of commitment to FoE/Amnesty*

	New (%)	Veteran (%)	Lapsed (%)	Weighted FoE sample (%)	Amnesty (%)
Very committed	10.5	18.2	10.9	17.0	23.2
Fairly committed	60.4	65.2	34.9	64.4	63.8
Not very committed	24.6	14.7	39.5	16.3	11.3
Not at all committed	2.6	1.6	14.3	1.8	0.3

Table 5.3 *'Do you intend to renew your subscription when it next becomes due?'*

	New (%)	Veteran (%)	Weighted FoE sample (%)	Amnesty (%)	NFU Countryside (%)
Definitely	57.8	68.2	66.5	78.2	59.8
Probably	30.7	24.2	25.2	17.7	28.3

Amnesty/FoE members exhibit a relatively high sense of political efficacy. As Rosenstone and Hansen (1993: 15) highlight, (political) efficacy has two aspects: it provides 'a sense of personal competence in one's ability to understand politics and to participate in politics (what political scientists call internal efficacy), as well as a sense that one's political activities can influence what the government actually does (external efficacy)'.

When asked whether their support had an effect on FoE/Amnesty's ability to 'protect the environment/protect human rights', 71 per cent, 75 per cent, and 70 per cent of new, veteran, and Amnesty members respectively believed that their support had an effect (see Table 5.4). Indeed, the most interesting data relates to the exceptionally few respondents who felt that their participation had no effect (7 per cent FoE veterans, 7 per cent of Amnesty members). These responses suggest that public interest group members do not accept the free-riding premise.

Table 5.4 *'What effect do you think your support of FoE/Amnesty International has on that organization's ability to protect the environment/protect human rights?'*

	New (%)	Veteran (%)	Lapsed (%)	Weighted FoE sample (%)	Amnesty (%)
A significant effect	9.9	14.1	7.1	13.5	14.9
A noticeable effect	61.3	60.6	39.5	60.7	54.7
No Effect	11.5	7.3	26.1	8.0	6.9
Don't Know	16.0	16.8	25.2	16.7	21.3

Olson's assumption is that members do not join in support of collective benefit, but our basic evidence is that members find support for campaigning more appealing than he suggests. There is apparent support for collective goals as an inducement to membership (see Table 5.5).

Table 5.5 *Responses to a potential change in the balance of group benefits offered by FoE*

(a) Would you rejoin/remain a member if FoE reduced direct member services?

	New (%)	Veteran (%)	Lapsed (%)	Weighted FoE sample (%)
Yes/definitely/probably	77.6	87.8	26.0	86.2
Definitely not	2.6	1.9	16.8	2.0

(b) Would you rejoin/remain a member if FoE reduced core campaigning activities?

	New (%)	Veteran (%)	Lapsed (%)	Weighted FoE sample (%)
Yes/definitely/probably	32.9	31.3	5.9	31.5
Definitely not	27.8	32.3	54.6	31.6

Table 5.5 shows that the overwhelming majority of new (78 per cent) and veteran (88 per cent) FoE members said that they would remain in membership if FoE reduced direct member services in favour of core campaigning, while approximately one-third of new and veteran members said that they would remain in membership if FoE's core campaigning activities were reduced. (Only 6 per cent of lapsed members would be persuaded to rejoin if core campaigning was reduced.)

Selective material and collective incentives

This section contrasts the relative importance/unimportance of selective *material* incentives to NFU Countryside/FoE and Amnesty

Table 5.6 *'A number of reasons why people might join NFU Countryside are listed below. Please indicate how important a role each reason played in your decision to join.'*

Reasons for joining	Very Important (%)	Important (%)	Not very Important (%)	Played no role whatsoever (%)
(a) As a member I felt I could join like-minded people in protecting the countryside	12.3	35.6	15.1	27.4
(b) To keep me informed about countryside and environmental issues	24.7	45.7	11.4	11.0
(c) I felt that NFU Countryside would provide me with information to make the best use of my smallholding	19.6	22.8	20.5	26.5
(d) I appreciated the way the NFU assists farmers and wanted something similar	18.7	30.6	19.2	23.3
(e) I wanted access to competitive insurance and other discounts	63.5	15.1	9.1	9.6
(f) I wanted sound professional advice on legal and other matters	26.9	34.2	13.7	16.0
(g) I wanted to express my belief in the importance of the countryside	15.5	33.8	17.8	22.4
(h) Other reasons which were important	2.3	4.1	4.1	18.7

members. It is doubtful if anyone would join FoE solely for the magazine *Earth Matters*, but large numbers of individuals *do* join NFU Countryside solely for the selective incentive of cheap insurance. For example, when asked the open question – *Why did you join NFU Countryside? (Were there any particularly important reasons or specific events which encouraged you to join?)* – 55 per cent of respondents mentioned the insurance services offered to members as an important reason for joining. When pressed on the importance of the selective benefits NFU Countryside members placed even greater stress on them (see Table 5.6): 79 per cent stated that 'access to competitive insurance and other discounts' was very important/important; 61 per cent said that 'wanted sound professional advice on legal and other matters' was very important/important; and 70 per cent stated that being kept 'informed about countryside and environmental issues was very important/important.

As argued in chapter 3, there is a higher level of public interest group membership than is predicted by Olson's work. The 'explanation' of this development rests on replacing the strict public choice paradigm used by Olson with an acceptance that those in membership may not be seeking selective material incentives, but might get a selective reward from playing a part in supporting a cause. For example, Rothenberg (1988: 1137) found that 72 per cent of Common Cause members joined for *purposive benefits* (e.g. 'supports general goals, issues, or efforts'; 'keeps government honest and fair'; 'supports public interest and the common good' etc.), 26 per cent joined for *selective benefits* (collective or divisible) (e.g. 'Provides political information'; 'offers a chance for personal political activity' etc.) and 1 per cent joined for *solidary benefits* (e.g. 'Family or friends belong', 'Joining is part of lifestyle – social reasons' etc.). In this perspective if one favours environmental ends, and the costs are low and bearable, then it would be perverse not to give support (see above, Table 3.1).

In response to an opportunity on the questionnaires to write in reasons for joining the groups, there were occasional responses that could be interpreted as 'selective and material' but very few:

'They have been very helpful with all the information I needed.' (FoE)

'Information for school children.' (FoE)

'I was taking a BA in Geography ... I saw membership as a helpful source of information in my studies.' (FoE lapsed)

'I teach Geography and Environmental Science in a comprehensive school and it is obviously good to be able to say I belong to various organizations and to have access to information posters.' (FoE lapsed)

Other answers have a hint of selective but non-material qualities. For example: 'a feeling of being welcome among concerned, like-minded people' (FoE) echoes the solidaristic incentive identified by Clark and Wilson (1961): 'Saw it as a means of meeting people with similar views at a lonely time in my life' (FoE lapsed). Most FoE and Amnesty answers are clearly and consciously in support of collective action – though the 'feelgood' factor obtained from this support is itself a selective reward. Two main trends intertwine. One is the 'stand up and be counted' instinct that recognizes the clout of big membership numbers, and the other is that support for FoE and Amnesty is a vicarious activity for those who are more passive:

'They actually take action which is impossible for an individual.' (FoE)

'The need of doing something, however little, to support a worthwhile group.' (FoE)

'FoE is a brand which means: we care and we are doing something. I wanted to support others who are prepared to be more active than me.' (FoE)

'It's so important to stand up and be counted on such an important issue. Groups such as FoE must be supported and given a high profile.' (FoE)

'I joined to be a number, simply to show governmental bodies that ordinary people care.' (FoE)

'I believe the more members an organization such as FoE has the more weight it carries with government, industry, etc.' (FoE)

'I feel I must do something to save the environment and giving Friends of the Earth the resources to do this seems the best way at present.' (FoE)

'Their power as a lobbyist group rests – to some extent – on the size of their membership.' (FoE)

'Joining gives the equivalent of a feelgood factor – a feeling that I've done something positive.' (FoE)

'AI was the one organization concerned with human rights internationally. I wished to be counted among its supporters and believed that the size of its membership was important in the work it was doing.' (Amnesty)

'AI enables one to do something, however small.' (Amnesty)

'Having seen the adverts I couldn't not join. I knew I would have little time to give, but felt that being a number might help a little.' (Amnesty)

Even lapsed FoE members echoed the same concerns:

'I feel that everyone has a duty to be "counted".' (FoE lapsed)

The tenor of NFU Countryside responses were starkly different. When asked about the relative importance of different considerations in joining NFU Countryside the two most important answers were: 'I wanted access to competitive insurance and other discounts' (64 per cent said this was 'very important'); 'I wanted sound professional advice on legal and other matters' (27 per cent said this was 'very important'). There was far more of the Olson self-interest. Members were frank about their motivation. Comments written in as 'other reasons' underscored the importance of selective material factors:

'To see if it was of any use to me.'

'To take advantage of NFU insurance schemes.'

'Because I bought a tractor and had to join the NFU to get tractor insurance.'

'Membership was requirement for car insurance.'

'To gain entitlement to less expensive insurance.'

'To obtain help and guidance on countryside matters. Grants available, NFU insurance.'

Such replies were overwhelmingly dominant. Insurance was volunteered by most members as a reason for joining. There were a few other more collective motivations expressed:

'I just felt I wanted, in a small way, to support the NFU.'

'Sense of belonging to a society concerned with our well being – a guide to cultural countryside living.'

'I am an agricultural adviser in private practice and wished to support the industry and NFU from within.'

'I care about the countryside.'

A separate question sought a comparison of four main reasons for joining NFU Countryside (see Table 5.7).

Table 5.7 *Most important reason for joining NFU Countryside*

	Most important reason (%)
(a) To obtain useful information	25.1
(b) To obtain discounts and cost savings	6.8
(c) I wanted to express my belief in the importance of countryside	4.6
(d) Access to NFU Mutual Insurance	58.0

How? Does the route into the organization make a difference?

Godwin (1988) has argued that the *mode of recruitment* is significant. He distinguishes between two basic types of recruits – social network and direct marketing. Those who joined because of friends or who were given a membership were classified as social network recruits. Individuals who responded to a direct-mail solicitation or other advertisement were placed in the direct-marketing category (Godwin, 1988: 51). Godwin's hypothesis is that direct-marketing recruits (direct mail) were less committed to their political groups than members recruited via social ties (personal contacts, local networks, word of mouth etc.). Godwin found that direct-marketing recruits stayed in the group for a shorter amount of time, they participated less in group activities, and had less attachment to the group – even though they were as likely to see the group as being effective.

Table 5.8 *'How did you first join FoE/Amnesty?'*

	Weighted FoE sample (%)	FoE lapsers (%)	Amnesty (%)
(a) I got a membership application form from a friend, relative or work colleague	8.8	10.5	7.2
(b) I contacted the Friends of the Earth office after seeing a press/media campaign (FoE); I responded to a press advertisement (AMN)	23.6	30.3	28.2
(c) I responded to a membership appeal I received through the post	22.6	22.7	18.5
(d) I joined at a national Friends of the Earth/ Amnesty meeting/conference/event	0.8	0.8	1.7
(e) I became involved at a local Friends of the Earth/Amnesty group meeting	4.8	2.1	3.6
(f) I joined Friends of the Earth/Amnesty after filling in an application form from a leaflet dispenser	16.3	12.6	9.9
(g) I wrote to Friends of the Earth/Amnesty International for information on environmental/ human rights issues	8.2	8.0	4.7
(h) I received my membership as a gift	1.7	4.2	0.0
(i) Rock concert (U2, Peter Gabriel)	0.0	0.0	8.3
(j) Other	11.1	5.9	10.8
(k) Missing	2.1	2.9	7.2

Our data (see Table 5.8) showed that Amnesty/FoE members tend to join as a result of group stimulation. This is in contrast to the membership of the Green and Conservative parties which have many members who joined via social network ties. For example, 23 per cent and 22 per cent of Conservative Party members respectively joined as a result of social contacts, and family contacts (Whiteley *et al.* 1994: 78). Similarly, 37 per cent of Green Party members said that: 'Talking to a Green Party member was Decisive/Very Important/Important factor in their decision to join' (Rüdig et al., 1992: 58).

Following Godwin's distinction our Amnesty International sample has been broken down into direct marketing recruits (DMRs) and social network recruits (SNRs). The first thing which is noticeable is the sub-group sizes. Of the 362 respondents who can be classified by these criteria only 45 (12 per cent) were in the social network category and 235 (65 per cent) were direct marketing. FoE's 1992 survey also found a high number of DMRs: 51 per cent of MSOs and 31 per cent of members joined via an appeal through the post, an advertisement, or leaflet. *These sorts of organizations are largely constituted by DMRs*: ironically this makes comparison difficult in Godwin's terms because of the low proportion of social network entrants. But the difference we found was trivial. Both the DMRs and SNRs were committed to Amnesty: 91 per cent of SNRs and 86 per cent of DMRs stated that they were committed members of the group, and 98 per cent and 96 per cent of SNRs and DMRs respectively said that they intended to renew their subscription. Somewhat unsurprisingly 44 per cent of SNRs and 9 per cent of DMRs said that talking to a member of the groups was important in their initial decision to join. Both sets of recruits differed little in their perceived political efficacy, with 64 per cent of SNRs and 70 per cent of DMRs believing that their support had a significant or noticeable effect. With regard to the most effective type of activities, gender, education level, class, and income, there is very little difference between the two sets of respondents. The major difference between the sets appears to be in the level of their 'active' participation in group activities – which Godwin also noted: 27 per cent of SNRs were members of a local Amnesty group and had taken part in an *Urgent Action Scheme*, while only 13 per cent and 9 per cent of DMRs respectively were members of a local group or had taken part in an Urgent Action Scheme.

Socio-demographically and attitudinally similarities rather than differences prevail. The majority of both groupings are female, educated to degree level, and have household incomes over £20,000.

Attitudinally the similarity continued: for example, 91 per cent of SNRs and 86 per cent of DMRs said that they were 'very/fairly committed to Amnesty'; 98 per cent of SNRs and 96 per cent of DMRs said that they 'intended to renew their subscription when it next comes due'; and 64 per cent of SNRs and 70 per cent of DMRs believed that their support for Amnesty had 'a significant or noticeable effect on that organization's ability to protect human rights'.

Post-materialism

Inglehart's theory of post-material value change utilizes Maslow's 'theory of human motivation' which posits that individuals have a set of *needs* for survival that they rank in order of importance and which they act to satisfy. The needs in the hierarchy range from 'physiological' and 'safety' needs to 'esteem' or 'self-actualization' needs. Maslow (1943: 136) argued that: 'The clear emergence of these needs (self-actualization) rests upon prior satisfaction of the physiological, safety, love and esteem needs.' Inglehart (1995: 64) maintains that 'self-realization' is related to post-materialism and that the possession of post-material values has an actual causal effect on behaviour. Post-materialism rests on two main hypothesis: a scarcity hypothesis which asserts that 'one places greatest subjective value on those things that are in relatively short supply', and a socialization hypothesis which asserts that 'one's basic values reflect the conditions that prevail during one's pre-adult years' (Inglehart, 1990: 56, quoted in Scarbrough, 1995: 124). For example, Inglehart (1995: 69) argues that:

Postmaterialists emphasise careers, rather than child rearing, as central to a fulfilling life for women. Though most Materialists believed that a woman needs children in order to be fulfilled, two-thirds of Postmaterialists in our samples reject this view ... societies with relatively high proportion of Postmaterialists tend to have much lower birth rates than other societies.

With regard to political participation Inglehart translated these needs into two fundamental categories – material and post-material. Inglehart's main thesis was that those who possessed more material needs would be less likely to participate in politics than the post-materialists. Bluntly stated, the poorest sections of society will tend to value the more 'basic' material needs, while the most affluent – who have satisfied these basic needs – will value the more aesthetic post-material needs (at the top end of the hierarchy). As Scarbrough (1995: 12) argues: 'The shift to post-materialism is thus a major force in the rise of a "new class" in Western society: "a stratum of highly educated and well-paid young technocrats, who take an adversarial stance toward their society" [*elite-challenging*], in sharp contrast with the "politically conservative" [*elite-directed*] outlook characteristics of earlier highly educated groups' (Inglehart, 1990: 67, 332).

In looking at participation those with post-material values should tend to vote for political parties which were addressing a wider and more holistic political agenda – e.g. ecology parties. Post-materialists would also be expected to have higher SES, and would be more likely to participate in the 'unconventional' political activities associated with NSM organizations. Drawing on data from *Eurobarometer* surveys (17, 22, and 25) which assess the support and potential membership of NSMs, Inglehart (1990: 54–60) found that:

In each of the twelve nations, post-materialists are far more likely to be members or potential members of these movements than are materialists. In any given country, post-materialists are at least twice as likely to be members as materialists, and the ratio is often three to one, or even four or five to one ... Materialist/post-materialist values priorities prove to be our strongest predictor in both the ecology movement and the anti-war movement across these twelve nations.

Our questionnaires gave respondents the opportunity to rank the preferences from four basic materialist/post-materialist value indicators. Table 5.9 shows the first preferences of the respondents to the question: *There is a lot of debate these days about what this country's goals should be for the next ten or fifteen years. Here is a list of some of the goals that different people say should be given top priority. We ask you to choose the two goals that are most important to you. Place the number '1' beside the goal that appears to you most important and '2' beside the second most important goal.*

Post-materialists would tend to select a 'self-expression or quality of life option', in this case either (b) or (d), and materialists would select an 'economic and physical security option' – (a) or (c) (Inglehart, 1990: 47). In all categories of membership the vast majority of FoE and Amnesty members selected a post-material option. Veteran, new, and lapsed members of FoE gave almost identical responses to the post-material question – 70 per cent of new and veteran, and 69 per cent of lapsed members selected the post-material preferences – thus lapsers are not distinctive in post-material terms. Seventy per cent of FoE members and 79 per cent of Amnesty members selected post-material options as their first choice. Indeed, the NFU Countryside provides a remarkably clear-cut contrast to the FoE and Amnesty position. Its members tend to be materialist in their value orientations – compatible with the finding above that they join for material selective incentives such as cheap insurance.

Table 5.9 *Respondents, 'first' post-material preferences*

Preference	Weighted FoE sample (%)	Amnesty (%)	NFU Countryside (%)
(a) Maintaining order in the nation	22.3%	13.5%	47.0%
(b) Giving people more say in important government decisions	47.4%	39.2%	19.6%
(c) Fighting rising prices	4.0%	2.8%	13.2%
(d) Protecting freedom of speech	22.2%	39.8%	10.0%

Those in our samples who feel closest to the Conservative Party tend to belong to the materialist category, while those who feel closest to the Green, Labour and Liberal Democrat parties tend to be more post-materialist. But when we look at class and post-materialism a less predictable picture emerges. Those respondents who identified themselves as working class are as 'post-materialist' as the middle class. Seventy-six per cent of middle-class respondents selected post-material options as their first choice, while 79 per cent of the working class selected the same options.

However, while our data has found strong support for post-material values among Amnesty and FoE members, Nas (1995: 288) is critical of the straightforward assumption that 'environmental issues are mere reflections of post-materialist concerns'. First, environmental problems may not be a higher-order abstract concern. The rise in environmental concern may just as easily be a result of environmental deterioration as the growth of post-materialism. Thus the salience of the issue may actually be strongly correlated with the objective condition rather than post-materialist orientation. It could be argued that increasing pollution affects an increasing proportion of the population *personally*, consequently it may not be seen as an abstract concern. 'These arguments are not mutually exclusive.' There is evidence that environmental consciousness springs from both post-material attitudes and adverse environmental conditions (see Rohrschneider, 1988).

Secondly, Nas (1995: 288–9) points out that it is worth noting that although post-materialists tend to give a relatively high priority to non-material goals, 'this does not automatically imply that material goals are unimportant to them. The term "priority" says nothing about relative strength of preferences ... it is not self-evident what will happen when post-materialists are confronted with a choice between a cleaner environment and their own mobility, such as losing their car or giving up their holiday flight to the tropics.'

Rüdig *et al.* (1993: 33) argued that post-material values are: 'only weakly correlated with green voting. There are lots of non-post-materialists voting green, and lots of post-materialists not voting green ... There is no statistically significant relationship between left-wing political orientation and green voting.' Our Amnesty data suggests that those who 'feel closest' to the Green Party are predominantly post-materialists, and that there is a relationship between a left-wing orientation and post-material values. Those who 'feel closest' to the Labour and Liberal parties appear to share post-material values (see Table 5.10).

Table 5.10 *Amnesty respondents, 'party closest to' and 'class': first material/post-material preferences*

Preference	Conservative (%)	Green (%)	Labour (%)	Liberal Democrat (%)	Working class (%)	Middle class (%)
(a) Maintaining order in the nation	47.6	9.5	9.0	13.4	10.7	16.3
(b) Giving people more say in important government decisions	4.8	61.9	40.7	45.4	50.0	39.1
(c) Fighting rising prices	4.8	0.0	2.1	3.1	7.7	2.7
(d) Protecting freedom of speech	40.0	23.8	47.0	34.0	28.5	36.4

Lapsing

The main perspective in the group literature on the issue of lapsing is provided by Rothenberg (1992: 21) in his work on membership retention. He says that standard models imply that members should remain in a group more or less for ever, but the experiential approach that he advanced lays stress on the rejoining decision. He suggests that joining can be a sensible part of a decision process: the member joins to discover more, and depending on that experience he or she will then stay or leave.

Table 5.11 *Main categories of response to* open *question: 'Why did you not renew your membership of FoE? (Were there any particularly important reasons or specific events which affected your decision not to rejoin?)'*

Reason for non-renewal	FoE lapsers (%)
Prefer to support other organizations/not happy with FoE	26.5%
Lack of money	33.2%
Inertia/Apathy/No time	6.7%
Not aware membership lapsed/did not receive renewal form	2.5%

In terms of general characteristics Veteran and Lapsed members of FoE do not fall into distinct categories. Veteran members are distinguished by being *more positive* in their opinions of FoE, but generally lapsed members are also surprisingly supportive: 21 per cent of 'lapsers' considered themselves to be members of FoE. Some of them had made a *one-off donation* to FoE in the previous year and considered their contribution to be a membership renewal payment (in some cases the contribution was actually greater than the cost of 'joining'). Others had simply forgotten to pay their subscriptions, and others still felt that they were members even though they had not made a financial contribution in the last 12 months. When asked an open question why they had not renewed their subscription (see Table 5.11) 27 per cent of 'lapsers' claimed that they were either dis-

satisfied with FoE or that they preferred to support other organizations; 33 per cent claimed that it was because of a lack of resources,
but lapsers tend to be relatively well-off in terms of household
income. Cost could be being used as a rationalization of other reasons.

This 'accidental' lapsing also emerged as important in the
Amnesty International study of the phenomenon in 1993. Their
own report stated that: 'Generally members, paying subscriptions
by Standing Order or Direct Debit (automated payment services)
who have been classed as "lapsed" have not forgotten to renew their
membership but (due to administrative weakness) ... have been
listed as such.' Looking at the four sub-types of FoE membership
(see Table 5.12), the explanation of 'lapsing' as being a response to
dissatisfaction seems weak. Over three-quarters still think that FoE
is an effective organization.

Table 5.12 *FoE sub-group differences*

	FoE new (%)	FoE veteran (%)	FoE lapser (%)	Weighted FoE sample (%)
Very/fairly committed	70.9	83.4	45.8	81.4
Renew subscription:				
definitely/probably	88.5	92.4	n.a.	91.7
Effect of personal support:				
significant/noticeable	71.2	74.7	46.6	74.2
Stay in membership if FoE				
reduced direct members				
services: definitely/probably	77.6	87.8	n.a.	86.2
How effective is FoE:				
very/fairly	94.2	92.9	77.3	93.2
Female	61.0	59.0	72.3	59.3
Degree/postgraduate	45.3	56.0	38.6	54.2
Self-perceived class				
position: middle class	68.1	74.7	64.7	73.7

Our FoE and Amnesty data confirm the flavour of the findings of
the 1993 Amnesty survey of 'lapsers' that: 'Only a handful, 4 per
cent, listed principled reasons such as disenchantment with
Amnesty or disagreement on a particular policy (e.g. adopting an

IRA terrorist as a Prisoner of Conscience).' Some specific policy issues prompted non-renewal. These might be grouped as 'policy non-renewals'. For example, there were responses such as:

1   Decision to accept IRA terrorists as prisoners of conscience.
2   Decision to accept US and UK soldiers as prisoners of conscience during the Gulf War.
3   Decision not to accept Mordechau Vanum case on the grounds that he was not a prisoner of conscience.
4   AI's commitment to the abolition of the death penalty regardless of circumstances.
5   AI's condemnation of the detention of illegal immigrants. Immigration (activity) seen as 'unrealistic, expensive and a waste of time'.
6   AI silence on question of abortion and family planning purges in the Third World.
7   AI losing sight of original ideals. Disbelief that AI remained politically neutral: considered left wing. AI associated with politically minded groups.

There were also 'non-renewals because of organizational features':

1   Fund raising methods seen as too aggressive.
2   Amounts suggested on donation box on forms considered too high.
3   'Shock tactic': material used in advertisements and the members' magazine very disturbing.
4   Members' journal uninformative with more material given to the press than members.
5   Frequency of appeals annoyed some members. View that there were too many appeals to be cost effective. Too much being spent on administration and paper.
6   Objection to reciprocal mail (when Amnesty shared contact names with other organizations) and objection to phone contact.
7   Cheques not cashed quickly enough. Poor recording of standing orders and direct debit. (List of points based on AI Lapsed Members Survey, 1993.)

Our data appears to suggest that the technique of membership payment is significant (see Table 5.13). If the group can 'steer' a member into one practice of payment rather than another it can increase the probability of retaining that individual membership. While attitudinally 'veteran' members of FoE were not that different from 'lapsed' members, in terms of whether they paid by 'direct debit'[1] or a conscious annual subscription, there was a clear distinction. In terms of income the categories are very similar. However, those not using direct debit were much more likely to allow their membership to lapse: 31 per cent of veterans were on direct debit (automatic payment) while only 6 per cent of lapsers used this method of payment.

Table 5.13 *FoE membership retention: method of payment*

	FoE veteran (%)	FoE lapser (%)
Direct debit users	31.0	5.9
Income +£30,000	22.5	22.2

Some reasons are inevitably entirely idiosyncratic. In at least one way it is easier to be dissatisfied in selective material incentive based organizations. If the attraction is economic benefit then experience can prove disappointing for a small minority. For example, a handful of NFU Countryside members gave negative responses:

'Waste of time/money and insurance company unsatisfactory.'

'The insurance premium is no longer cheaper than other companies.'

'NFU insurance could only offer third party cover for our ten sheep at the same rate as if we had 100.'

'Expectations raised in advert do not match facts. The organization comes over as some information on countryside matters plus a plethora of sales and in particular insurance pressure selling. No voice to express feelings or opinions, since members have no vote within NFU.'

If the attraction is economic saving then failure is simpler and starker to perceive than where the decision is judgemental about political effectiveness or about the satisfaction in expressing support for a cause.

Levels of activism

Rothenberg (1992: 127) has suggested that as well as a decision whether to stay/leave membership, there is an important decision on levels of activism. He argues that: 'Contributors can elect to write their dues check and stay in the rank and file, or they can choose a higher level of involvement.' Rothenberg discerns two types of above-normal activism:

1 *Chequebook activists* – who donate additional monies; and
2 *Temporal activists* – who volunteer their time.

In our analysis of Amnesty we also distinguish two further types of supporter:

3 *Chequebook supporters* – who give the basic financial contribution; and
4 *Super activists* – who give more than the minimum in terms of time *and* money (see Table 5.14).[2]

Table 5.14 *Active participation by Amnesty International members*

Activity	Chequebook supporters (%)	Temporal activists (%)	Chequebook activists (%)	Super activists (%)
Member of a local Amnesty group	0.0	19.6	0.0	26.3
Took part in urgent action scheme	0.0	11.8	0.0	24.0
Took part in letter writing campaigns	0.0	74.5	0.0	81.0

The outcome of our analysis appear largely predictable, but not uninteresting. While chequebook supporters exhibit high levels of commitment to the group they are, nevertheless, the least committed (and the least active) members: 67 per cent of chequebook supporters said they were very/fairly committed to Amnesty, compared with 82 per cent temporal activists, 86 per cent chequebook activists, and 93 per cent super activists. When we asked the open question – *Why did you join Amnesty International? (Where there any particularly important reasons or specific events which encouraged you to join?)*, 21 per cent of temporal activists and 15 per cent of super activists perceived membership as a means to actually do something to help fight for human rights, compared to 6 per cent of chequebook activists and 7 per cent of chequebook supporters. Super activists (fortunately for the credibility of the data) are much more likely to 'definitely rejoin' (84 per cent) than chequebook supporters (59 per cent). The minimalist chequebook supporters are less likely to link joining with a specific event or wish to 'do something' but answered more in terms of general concern/principles. Activist members exhibited higher levels of political efficacy: 75 per cent of super activists and 69 per cent of temporal activists believed that their support for Amnesty has a 'significant/noticeable effect', compared with 59 per cent of chequebook supporters. Rothenberg suggests that members with higher assessments of their associational and systemic efficacy are likely to be activists. Rothenberg (1992: 148) claims that 'results are more important than passion to Common Cause activists'. Our qualification is that while activists are likely to be confident about the beneficial effects of their participation so (to a lesser degree) are 'mere' supporters.

The overwhelming majority of respondents voted for either the Labour or Liberal Democratic parties. However, it was more surprising to find that 11 per cent of super activists voted for the Conservative Party at the 1992 general election. There was relatively little difference in the answers to the post-materialism dimensions, in self-assessed class position, and in household income levels. In terms of gender there was a more pronounced female bias: 63 per cent of chequebook supporters are male, but in the activist categories males are at best 40 per cent.

In summary we have found that most members of FoE and Amnesty exhibited high levels of commitment to these organizations; believed themselves to be politically efficacious; and joined for non-material (altruistic) reasons – selective *material* incentives were of limited value in explaining membership. Respondents joined because they believed they were backing a good group/cause.

Notes

1   This is the British form of automatic annual payment.

2   Our discussion here relates solely to Amnesty and we split out sample into four categories: chequebook activist = 19 per cent (n = 70); temporal activists = 14 per cent (n = 51); chequebook supporters = 13 per cent (n = 46); and super activists = 50 per cent (n = 179).

6

The contribution of marketing in an explanation of membership[1]

Marketing is thus built upon a paradox; it starts with the customer, is directed at the customer, but is fundamentally concerned with the satisfaction of the producer's own interests.

(Sackman, 1992; quoted in Wring, 1996: 3)

Marketing interest group membership

As shown in chapter 3 the existence of large 'individual membership' public interest groups casts a shadow over accounts which stress the difficulty of mobilizing and sustaining interest groups. Olson (1971 edn: 126) claimed that rational individuals will: '*not* voluntarily make any sacrifices to help their group attain its political (public or collective) objectives'. As Godwin and Mitchell (1982: 162) express it:

Olson's model predicts that persons will not join public interest groups such as Common Cause or Environmental Action that supply few selective incentives; yet millions of persons do join. While the argument can be made that joiners are a small proportion of those who value the collective goods, they represent millions of deviant cases for Olson's theory.

There is now a body of (largely North American) research which suggests that individuals join interest groups for a variety of reasons not part of the central Olson calculus (see Godwin, 1988; Moe, 1980a, 1980b; Rothenberg, 1992; Sabatier, 1992; Schlozman, *et al.*, 1995; and Walker, 1991). Even where individuals are seen as rational maximizers of their interests, the sorts of interests they wish to advance may not be as narrow as Olson set out. Empirical evidence has contradicted Olson's assumptions that public interest groups are fatally disadvantaged in terms of ease of mobilization *vis-à-vis* business groups which can offer selective material incentives to potential members.

Olson, and the early pluralists from whom he dissented, saw membership as individual-centred. Our preferred explanation of large public interest group support is based on a *group perspective*.[1] The *efficiency and nature of supply* is an important dynamic in the joining decision. Groups can by their marketing strategies alter the level of demand for membership, and given the low monetary costs of subscription to most public interest groups, joining/supporting decisions are below a 'threshold of (economic) rationality': i.e. below the 'cost' of exhaustive analysis. As accepted in mainstream economics, once basic needs are satisfied 'luxury' decisions may be taken on grounds other than financial self-interest (see Marsh, 1976: 270; and Barry, 1970: 40).

This chapter argues that groups attempt to develop membership by using professional marketing strategies which aim both to lower the perception of the economic cost of membership, and to crystallize the predisposed members' concerns. Groups offer a mixture of organizational and psychological strategies which help place the joining decision on individuals' agenda, and helps shape the decision-making process. Thus, modern, large-scale interest groups are the product of mail order marketing to an extent that has been inadequately recognized in Britain, though the perspective goes back at least a decade in the United States (see Berry, 1984: 82–6). Hansen (1985: 94) argues that 'supply generates its own demand' and we need to examine the issue of the role of *marketing* in group viability. *The direct mail appeal has replaced the begging bucket*: or as Deacon and Fenton argue: 'For charities the days of sitting back and rattling the tin are long gone' (*The Guardian*, 17 April 1996).

Groups influence decisions by potential members. As Dunleavy (1991: 48) observed: 'Financial muscle and organizational skills may help to advertise group activities and membership, and to persuade waverers.' The group role needs to be placed more centrally in the account of joining than it has been hitherto. Godwin's (1992: 309) discussion of the importance of direct marketing in politics helps explain the rapid growth of citizen action groups in the United States (e.g. Common Cause, the National Taxpayers' Union etc.).[3] (In chapter 7 we discuss the consequence of the growth in direct-mail and direct marketing for participatory democracy.)

The groups we discuss are essentially *mail order* groups (Mundo, 1992: 18)[4] where members have little (if any) influence by way of a policy-making role: in terms of participation these are *low demand* organizations.[5] It is an implicit aim of this chapter to stress the gap between the nature of these organizations and the social movements (/organizations) discussed in chapter 2. Furthermore, members are not the 'holistic' environmentalists of an ecological type. Groups have found success in single species, niche-market differentiation. 'Saving the cuddly thing' is a more effective marketing pitch than a general appeal about saving the environment. Even general environmental groups tend to develop *de facto* main strands, if not single strands – in the 1980s FoE was very much a 'rain forest' group. So the mobilized member is often 'sucked in' precisely because of the *particularist* rather than the *general* environmental appeal.

Large groups, as studied in this volume, are particularly important because it is such bodies that yield the large numbers of 'members' that appear so impressive in discussions about the decline of party and the creation of alternative modes of participation. But size is linked to recruitment strategy: Johnson (1995: 15) shows that the direct mail strategy in the United States is particularly associated with the largest groups. Six out of the seven largest groups in his sample of environment groups used direct mail. Cohen (1995: 176) notes that Amnesty USA sends out 11 million direct mail letters every year.

Large public interest groups must retain what Hayes calls 'checkbook' members to justify their marketing and recruitment. He argues (1986: 134) that some groups guarantee their survival by building up this type of cash-cow membership category:

The vast majority of new groups in recent years fail to conform to the traditional conception of a mass-based membership group. Many of these groups are staff organizations, lacking any real membership base. Others have a membership consisting of 'checkbook' members, affiliated with the group only by virtue of monetary contributions and lacking any face-to-face contacts with other members in a common setting.

It is possible to mischaracterize these organizations if they are seen as mass participatory arenas. Instead they are part of the funding of protest rather than a means for much individual participation.

Starting points for a supply-side perspective

There are six broad preliminary points that underpin this argument.

1 Like those he criticized, Olson regarded membership as the product of individual choice: more needs to be known about the supply side. Green and Shapiro (1994: 15) claim that an important element in the rational choice approach is that the relevant (utility) maximizing agents are individuals. In fact, groups manipulate the individual choice in significant ways. For example, an important factor in joining a group is that a well-publicized group already exists and makes itself accessible to new members (Hansen, 1985: 93).

2 There is such variety of interest groups, and such diversity in the reasons for joining or not, that we are sceptical of the feasibility of *the* membership explanation. Dunleavy (1991: 54–5) distinguishes between *exogenous* groups, whose members share an 'identity set' – i.e. some objective characteristic such as a professional status; and *endogenous* groups which sees the combining of 'likeminded people' where the potential members are spread through the society – i.e. 'cause groups' such as Amnesty International. *Exogenous* groups are able to make direct contact with their potential membership in a targeted way – all doctors are potential members of the British Medical Association. Here we discuss the marketing approach associated with *endogenous* groups that have no *simple* way to identify their potential constituencies. For these groups, direct mail and heavy press advertising are important ways of locating potential supporters.

3 Is the membership phenomenon the same for the strongly political group as for the politically marginal? Membership of the Environmental Transport Association[6] (ETA) is a more deliberate political act than that of the Automobile Association (AA) (where the main membership incentive is roadside assistance) – though the former is less influential in the policy process. Recruitment for the strongly political group might well be different for the politically marginal group. Good marketing might work to bring in members for the AA but not for the ETA where enthusiasm for the cause might be more vital.

4 The decision by the individual can be *less* rigorous and complex than elaborated in a developed public choice perspective. In the extreme there is the 'flag day response' where the 'cost' of con-

tributing is smaller than the social/emotional costs of refusing. More importantly for this book, rational choice approaches exaggerate the costs of joining which can be manipulated downwards by group recruitment strategies (see Godwin, 1988; and Rothenberg, 1992). 'Soft', non-economic, solidary, expressive, and purposive benefits are central to an explanation of joining such groups (Opp, 1986). However, those groups which are best at getting members via marketing strategies may aggravate turnover problems. Their marketing efforts may bring in *marginal* memberships less likely to stay in the group. Members who actively seek the group are likely to be less fickle in their allegiance.

5 Groups are active on their own account in a way not set out by Olson. They are organized to stimulate applications and not simply to respond to spontaneous joining. US data shows that Greenpeace spent 23 per cent of its income on fund-raising (and a further 4 per cent on internal management) while the Sierra Club spent 22 per cent (Bosso, 1995: 111). Cohen (1995) records that membership organizations like Amnesty devote large resources to appeals. He set out the three most common techniques: direct mailing; advertising in newspapers and magazines; and loose leaf inserts in newspapers and magazines – 80–5 per cent of the annual revenue of Amnesty in Britain comes from these three methods. Since the methods are themselves expensive, the 'profitability' of the solicitations is marginal. Cohen (1995: 121) notes that in bad years such as 1993 these approaches barely break even. Amnesty USA used direct mailing as an experiment, and its newspaper advertising did not even (re-) cover the costs. (Twelve million direct mail letters were sent out to recruit 'new' Amnesty members with a take-up rate of only 0.5 per cent.) The industry rule of thumb is that a take-up rate of about 2 per cent for such appeals[7] is needed for direct mailing to be successful.

6 Finally there is a need to connect the discussion of joining interest groups to the literature of social psychology rather than the economic rationality. For example, Toch's *Social Psychology of Social Movements* describes bodies such as National Association for the Advancement of Colored People and the Anti-Digit Dialling League as social movements. However, his comments on social movements can be extended to the political science conception of interest group. In Toch's account there is an appreciation of the role groups play in *creating* membership. He quotes Herbert Blumer

who laid stress on the process of agitation *by* the group *on* the member:

The gaining of sympathizers rarely occurs through a mere combination of a pre-established appeal and a pre-established individual psychological bent on which it is brought to bear. Instead, the prospective sympathizer or member has to be *aroused*, *nurtured*, and *directed*, and the so-called appeal has to be *developed* and *adapted*. (Toch, 1965: 87)

Most of the political science work on group membership over the last three decades has been driven by the need to respond to Olson's version of the rational choice challenge. In the process, the potential contribution of political sociology and social psychology has been marginalized (Marsh, 1994). Unfortunately the rational choice orientation is often abstract and disconnected from the empirical world. At the very least, the sociological or psychological explanations may be particularly valuable in an explanation of the weakly political group (see Morris and Mueller, 1992, for a social psychological perspective).

Marketing membership

Olson's main argument is that groups will *not* emerge spontaneously merely because individuals share goals. This proposition is significant because it is a powerful criticism of 'complacent' pluralist arguments that beliefs/interests will automatically mobilize in the political process. However, we believe that while *actual* group membership reflects the sum of separate decisions about incentives as Olson predicts, the membership total is in part crafted by the group. The group can reduce the cost of membership and supply soft incentives as opposed to the material incentives on which Olson focused (see Opp, 1986). Modern large-scale campaigning groups which are the product of mail order marketing are – as argued in chapter 1 – essentially *protest businesses*.[8]

Toch (1965: 16) in anticipating this supply-side emphasis proposes that:

Social movements – *like other advertizers* – must show that they can respond to the needs of their clients. They must demonstrate their ability to furnish solutions which make it worth expending time, energy, and dedication. They must publicize offerings which people can find useful and desirable [emphasis added].

Group recruitment activities (e.g. canvassing, advertising, direct mail etc.) are aimed at attracting members who have not taken the initiative themselves. The group recruitment strategy reduces the cost (in terms of effort rather than cash) of joining. The (near) nine-fold increase in membership of the RSPB between 1971 and 1991 (98,000 in 1971, to 441,000 in 1981, to 852,00 in 1991) reflects the adoption, and success, of regular and high-profile press advertising and increased sophistication in recruitment, much more than a change in public attitudes. Group size or income is not simply a matter of the aggregation of individual decisions. Public interest groups have adopted the tactics of merchandizing organizations to stimulate the level of membership decisions.

Bosso (1995: 111) notes that the growth in mass membership environmentalism in the United States from the 1970s onward 'did not just happen simply because environmental issues became more important. It also was cultivated as part of a conscious effort by many environmental leaders to build member bases as another tool in their organizational and advocacy armamentarium.' Marketing is responsible for placing membership on the agenda for decision. For the group, not deciding about joining is as unhelpful as deciding not to join: large numbers of individuals would not join groups without the group initiative. They would come to no conclusion for or against. Joining opportunities at least give the group a chance of success.

As established in chapter 1 *protest businesses'* staff are recruited to membership departments to promote and market a group on the basis of a proven professional marketing track record – not necessarily because they are committed to the cause. For example, an advertisement (in *Scotland on Sunday*, 10 September 1995) for the post of 'Membership Recruitment Co-ordinator' at the National Trust for Scotland stated that it was 'part of the Trust's marketing strategy to increase our level and quality of membership recruitment'. Applicants were 'likely to have a background in sales'. In the United States, Thompson cites the job description for the Executive Director of the Sierra Club as one that 'might fit IBM's needs':

This person will have an outstanding financial background, and have achieved the position of CEO or have been groomed as the second in command ... Prior experience with an environmental organization is not required ... this position is not limited to a not-for-profit executive ... An advanced degree is highly desirable with a concentration in management or finance. (Thompson, 1985: 9, original emphasis)

The 'group-speak' has also adopted the language of business in recent years. For example, Johnson (1995: 19)[9] reports one group's entrepreneur as speaking from 'the literature on direct marketing, where the critical concepts are *investment* and the *lifetime value* of a member'. He quotes the respondent describing the group's objective in using (even though it lost many in its first year) direct mail campaigns to attract members as follows:

It is like a sort of capital investment. You put your money in and it comes out eventually, over time … You know, its fairly simple, and a lot more secure than a lot of the kinds of investments that people make all the time … like the stock market and that kind of stuff.

The large numbers of individuals 'mobilized' by mail order groups are primarily facilitating/permitting professionalized interest representation rather than themselves playing much of a role on the political process. Bosso (1995: 106) notes that professional interest representation is not cheap. He points out that a full-time staff with policy and managerial expertize is vastly more expensive to recruit and maintain than one comprised of the young volunteers who typified environmentalism in the early 1970s. The need to get funding from a broad support is then a consequence of the changed style of political intervention of the groups.

It is possible to argue that this interaction between individuals and groups is still within the scope of rational choice theory as the groups are 'rationally' attempting to alter the 'rational' decision of potential members. The breadth of rational choice approaches means that very different tendencies are covered. For example, Herbert Simon (1993: 49) has complained that his work has been fundamentally misinterpreted when it was lumped in with the work of economists. Whereas Theodore Lowi had subsumed Simon's work with that of the rational choice school, Simon protested: 'Are … (my) books written so obscurely that Professor Lowi could not see that the rationality celebrated in them (if rationality is celebrated at all) is a weak, muddled, bounded rationality that is rejected out of hand by the economists who espouse public choice and neoclassical laissez faire theory?'

This chapter discusses certain issues also discussed in chapter 3 such as 'collective bads' and the importance of selective 'soft' incentives. In this chapter there is a different emphasis. Here we are stressing the way in which groups can manipulate perceptions of these matters. The next nine sections set out how the group can alter the decision frame of individuals.

Tapping existing predisposition

Godwin (1988: 54) argues that marketing taps, rather than creates, predisposition. If, as he argues, personal efficacy is a key explanation of political participation, then marketing 'reaches persons who already perceive themselves as efficacious, and moves them to political action'. The predisposed potential membership has usually been identified by the group by compiling profiles of past joiners and other market research tactics such as buying membership lists from other similar organizations. Cohen (1995: 119), notes that: 'even the most populist organizations only reach a restricted sector of the population: well educated, high socio-economic status, more liberal in their political views, already belonging to the "conscience constituency". Direct mailings are sent to targets identified by market research profiles or to lists exchanged or purchased from like-minded organizations (for example, supporters of environmental or development groups).' The RSPB follow up the 'warm names' left on the visitors' log book at their bird observation reserves as a source of new members. Groups will also, as Godwin (1992: 314) points out 'trade or sell' the names and addresses of those individuals who respond to one of their appeals. However, the success rate in terms of recruiting, even in carefully selected target audiences, is low. As Berry (1984: 83) points out, the group that buys a direct mail list will probably pre-test the list by contacting a sample and checking what percentage of those contacted respond.[10]

Cohen (1995: 170) discusses the difference between a 0.4 per cent and a 0.9 per cent response rate as being commercially vital. Indeed Greenpeace in the United States cut its mail appeals in half between 1990 and 1992 with a drop in revenue from that source from 62 per cent to 49 per cent. It has tried door-to-door canvassing and telemarketing and tried to get more income from its committed supporters and from corporations and foundations (Bosso, 1995: 117).

Within such tight margins groups have shown a considerable

degree of enterprise in recruiting new members. While Dunleavy (1991) argues that potential members of endogenous groups – such as FoE and Amnesty – are not immediately identifiable, the groups themselves have found that many potential members share some distinguishing characteristics. For example, our survey revealed that (in that year) FoE had recruited a significant number of members through an application form enclosed with the 'green' washing power ECOVER. The predisposition to join FoE is reflected in 'green' shopping decisions and the group exploited the 'green' factor exhibited by those who show concern about the environment in a variety of ways.

Thus, many endogenous groups (e.g. Greenpeace, Amnesty, FoE) do have clues in locating their potential members. These groups are *chasing the same chequebooks* and know *where to look for members* – or more accurately the members *they choose to have*. For example, a Greenpeace official said that they advertised in the *quality* press because 'the tabloids are not a good recruiting ground'. An Amnesty survey in 1993 found that their members read the quality press: 45 per cent *The Guardian*, 40 per cent *The Observer*, 36 per cent *The Independent* (Cohen, 1995: 200). Hence the *almost* identical socio-demographic profiles (and overlapping membership) of these organizations is no accident (see chapter 5). They are all targeting the identified constituency.

The argument that many public interest groups have successfully mobilized a *viable* proportion of 'their' predisposed public into membership since the 1960s in spite of Olsonian pessimism has two aspects. It means that large numbers are available to be recruited (unless Olson's free-riding objection holds), and low levels of success in terms of the percentage of predisposed potential members recruited can be significant in absolute numbers. *A relatively weak solution to the collective action problem can be sufficient.* For example, Greenpeace UK had around 440,000 supporters in 1992 but, arguably, there were millions more among the general population who were, at worst, not antagonistic towards the group. Viewed in this way, one might argue the group has recruited only a fraction of its target group: Olson is in part correct. Nevertheless 440,000 is a respectable critical mass which provides sufficient financial support and the appearance of representativeness. *Such bodies need only recruit a small proportion of the potential membership to have a viable group. A small percentage of a very large total can represent good business. A group can be 'big enough' to be policy relevant even though it is well below a maximum feasible size.*[11]The contribution of selective material incentives for public interest group membership may be about bringing a predisposed individual into membership – selective incentives (e.g. a free gift) may reduce members' entry costs. The important point is that the gift will be most appealing to those already interested in that area. The free gift tips the balance of costs and benefits in favour of joining. The group can thereby alter the nature of the decision made by the individual. This is, of course, an entirely rational manoeuvre by the group: but is not the sort of factor elaborated on by Olson.[12]

Creating predisposition

Groups also *shape predisposition*: concern at issues such as the depletion of the rain forest or the fate of whales can be generated. This involves drawing the potential member's attention to a 'problem' or capitalizing on natural disasters: groups must stimulate 'demand'. Greenpeace, for example, is regularly contacted by newspapers alerting them to advertising possibilities when an environmental story is about to be published. Thus, the group will advertise on the same page as a lead story and a picture of environmental damage or stricken wildlife. The RSPB Market Research Manager said that following the 'Braer' incident 'the RSPB ... recruited a few hundred new members at a ROI [return on investment] *better than* our normal advertisements'. Predisposition can be an artefact.

The sinking of the Braer oil tanker off the Shetland islands in 1993 was the type of public incident which immediately prompted group advertising. A selection of the appeals used (in most cases accompanied by pictures of stricken wildlife) included:

1 RSPB is fighting to keep Britain's birds and marine creatures safe from future disasters (RSPB).
2 The Price of Oil has Just Gone Up. Birds, seals and otters are all facing agonising deaths by drowning or poisoning (RSPCA/SSPCA).
3 As you read this, Greenpeace is lifting oil stricken birds and animals out of the seas around the wreck of the Braer ... we are using resources we would normally spend on campaigns which aim to stop this kind of thing happening in the first place (Greenpeace).
4 The Braer disaster ... has brought home the threats to Britain's wildlife in horrific fashion (FoE).

That the Braer incident turned out not to be the environmental disaster forecast is not relevant. The impact of the fear of disaster was useful for group recruitment. As argued below, Mitchell (1979) maintains that 'public bads' are more effective recruiting arguments than collective goods: there is an emotive power in the threat of deterioration that is lacking for more positive messages. A collective bad is of course like a collective good in the sense that the outcome is shared by all in an undifferentiated way whether or not one contributed. Logically one has the opportunity to free-ride activities to prevent the emergence of the collective bad just as one can free-ride any activity to secure any other collective good. The term has been devised, however, because there appears to be a difference in the way individuals regard that subset of collective goods termed collective bads. Through experience, groups believe that individuals are more likely to contribute to the prevention of a collective bad than the provision of a collective good.

Profiling: locating an attainable membership

Recruitment will be frustrated if the predisposed member cannot afford the subscription. Groups deliberately market themselves among those best able to afford support. Thus groups, like other well-organized businesses, generate *pertinent (marketing) information* (Wring, 1996) which enables them to identify the (shared) characteristics of an *attainable membership*. For some potential members the subscription is a reasonably serious decision, while for others it is not. It is obviously the latter category that interest groups pursue. As reported in more detail in chapter 4, the memberships of FoE and Amnesty are distinctive in socio-demographic terms (also see the data presented in Table 6.1).

One conclusion to draw from the data is that within groups such as Amnesty and FoE the middle class are over-represented because they have a greater sense of personal efficacy. Another factor however, is that membership reflects the *recruiting strategy* in a self-fulfilling fashion. Amnesty, for example, advertises in the quality press which is read by a middle-class readership which produces an organizational profile that suggests advertising in the quality press.[13] The relatively high disposable/discretionary income of readers of these papers means they can afford to be members. Without necessarily having a greater sense of efficacy, they can afford to indulge their

inclination. Groups have overwhelmingly middle-class memberships because this is the target audience. As Gerhard Wallmeyer, a
fund-raiser for Greenpeace in Germany said: 'The more educated
the addresses are the more chance we have that they become first
time donors' (*SPIEGEL Special*, November 1995).[14]

Table 6.1 *Level of education, income and class position*

(i) Highest educational qualification gained

	FoE new (%)	FoE veteran (%)	FoE lapsed (%)	Weighted FoE Sample (%)	Amnesty (%)
Degree	26.8	37.0	29.4	35.3	26.0
Postgraduate	18.5	19.0	9.2	18.9	26.5

(ii) Household income above £20,000

	FoE new (%)	FoE veteran (%)	FoE lapsed (%)	Weighted FoE Sample (%)	Amnesty (%)
£20,001–£30,000	22.4	21.2	19.7	21.4	19.6
£30,001–£40,000	11.5	10.3	13.0	10.5	13.8
Over £40,000	12.5	12.3	9.2	12.3	19.6

(iii) Self-perceived class position

	FoE new (%)	FoE veteran (%)	FoE lapsed (%)	Weighted FoE Sample (%)	Amnesty (%)
Working class	23.3	20.4	27.7	20.9	22.9
Middle class	68.1	74.7	64.7	73.7	71.3

Groups stimulate consideration of the decision to join

For highly political groups the potential member may well have the
motivation to take the trouble to track down the organization;
political parties have in the past been notoriously ill-organized to
admit new members. For members of weakly political groups – or

weakly motivated memberships of strongly political groups – the fact that the group takes the initiative in offering convenient access may be vital. As the conventional wisdom in another marketing lead business states, 'life insurance is *sold*, not bought': the level of (public interest group) membership then can be shaped by the group presenting *joining opportunities*.

Organizations such as Amnesty in Britain estimate that up to six or seven 'bites' (repeated exposures to adverts, inserts, or letters) are sometimes needed before a positive response comes and a member joins (Cohen, 1995). Johnson's (1995: 15) recent US study reports that 28 groups who responded to his questionnaire contacted 17.37 million non-members. Members received an average of 8.7 pieces of direct mail each year.

The implication of this industry assumption is that unless the group makes these sorts of joining opportunities available the predisposed members might not join. Thus the 'rational' decision sought by the group is more likely the more often it is faced. This argument might lead to the conclusion that groups should advertise continuously. In practice, however, the 'law of diminishing returns' leads to the costs of additional recruits exceeding the income gained. Large mail campaigns will probably not pay for themselves on a one-year basis. As noted above, the organization needs to keep the additional members several years to recoup the advertising 'investment'.[15]

Groups also have a strategy of giving lapsed members repeated opportunities to rejoin the organization before they admit defeat. 'Lapsing' can be converted to rejoining by group tenacity. It can also be avoided by getting the member to automatically rejoin without an annual decision. At Olson's level of abstraction, membership is membership is membership: it would not appear to matter which *technique* of payment is adopted. In fact, if the group can 'steer' a member into one practice of payment rather than another it can increase the probability of a 'good' decision. Our data (Table 5.13) showed that attitudinally 'veteran' members of FoE were not that different from 'lapsed' members; however, in terms of whether they paid by 'direct debit' or a conscious annual subscription, there was a clear distinction. Groups recognize the importance of 'moving' recruits on to an automatic payment scheme. For example the ETA waives a joining fee of £10 if direct debit is agreed.

Manipulating the perception of monetary and time costs

Membership departments exist to *supply the group* to a potential membership not to process spontaneous applications for membership. Groups seek to manipulate *down* the perceptions of the costs of membership and manipulate *up* the perceived consequences of not joining. They also seek to manipulate what Dunleavy (1991: 54) terms 'group identity'. The group will attempt to persuade the individual that he or she has a stake in the issue.

Groups reduce membership costs in several ways. FoE, for example, has four types of contributor: (i) 'members' who pay an annual supportership fee (with reductions for under 18s/students/OAPs/unwaged); (ii) supporters who are identified as having paid subscriptions regularly over a period of years; (iii) committed givers who are subscribers by direct debit contributions (various levels of commitment from £3 per month upwards); and (iv) donors who receive appeal literature and give one-off donations but who do not receive the regular correspondence sent to members. Group strategy is aimed at providing a variety of options for contribution. This process of segmenting types of members is also important in organizational maintenance. The groups can identify the *segments* with greatest financial potential, i.e. those who can be 'counted on' to respond to solicitations more regularly than annually. Segmentation allows the groups to identify their best customers.[16]

Groups appreciate the role of factors such as inconvenience of form-filling, poor personal organization and so forth in their recruitment and retention strategies. These 'costs' of joining are taken as seriously by potential members as the monetary subscription factor. For example, a 1992 Greenpeace direct mailing to lapsed members read: '*90% of people who fail to renew their subscription to Greenpeace do so* not *because they don't want to, but simply because they don't get round to it*' (emphasis added).

If members have to write an annual subscription cheque to stay in membership they have a more convenient opportunity to express dissatisfaction with the group (or simply not get around to rejoining) than those who pay via direct debit which reduces the visibility of membership for members: it removes the need for an annual decision. Many trades unions, for example, collect dues monthly directly from salaries and this helps to manipulate down the perception of the monetary costs of membership. Ten pounds per month automatically deducted from a salary is much less noticeable than a £120 once-a-year deduction, or the act of having to write a cheque for such an amount.

Olson saw public interest group participation as unlikely to be self-interestedly rational. However, if the membership cost is low, joining easy, and membership reinforces the self-image of the potential member, then membership is more likely than an economically based rational choice calculation would suggest (see Dunleavy, 1991; and Godwin, 1988). In the main, monetary costs are less significant than Olson's account tended to imply. In practice these *deliberately targeted* individuals do not need to make elaborate calculations about personal or group *pivotality* and efficacy in making decisions. Some will simply decide they will feel better having joined: the expressive benefit calculation need not involve the member in knowing anything other than their own welfare.

Olson's supporters have noted the low cost of participation in protest businesses to say that this confirms the Olson thesis. Hardin, for example, claims that environmentalists annually spend less on 'their apparently great cause than two pack a day smokers spend on cigarettes'. He says that the annual sum is 'a trivium.' If we accept this, then the argument can perhaps be reversed. If the sum is indeed trivial is it likely that potential members are free-riding on the grounds of cost saving? Free-riding is likely to be a reasonable response to significant investments but it is less reasonable for costs that are minimal.

Manipulating the value of collective goods

There is no need to assume that public interest group membership is explained by material self-interest or altruism: individuals selectively obtain gratification from 'other regarding' actions that *appear* altruistic. The reward is often a 'feelgood' or 'expressive' factor (Salisbury, 1969). One might like the confirmation of one's own beliefs given in membership newsletters. If individuals are 'self-regarding and instrumental' in their behaviour it does not follow that they free-ride (Dunleavy, 1991: 249). They might have self-regarding reasons for acting collectively. Margolis (1982: 103) argues that it is rational to allocate expenditure to a group which does not give selective material rewards if one allocates some portion of one's spending for social purposes. He argues that a donation to a charity is rational if the individual decides the social value of the (say) $10 if given to charity is greater than spending the $10 on himself.

Cohen (1995) notes that recruiters and fund-raisers talk of a hierarchy of needs (see Inglehart, 1977; and Maslow, 1943). The product they are selling, Cohen says, is a purchase for the relatively affluent who have satisfied other wants and now wish to build up their self-esteem. Economists would treat such decisions as luxury goods purchased rationally when other needs are satisfied. Olson assumes that there are identifiable collective political goals but that one would not invest one's own resources because these would not materially affect the probability of the goals being attained. In his 'economic man' perspective the rational result would be 'free-riding'. However, we suspect that the 'purchase' of group membership or the expressive act of making a financial contribution to a worthwhile cause might itself be valued. It is impossible to free-ride this: to free-ride the cost is to free-ride what was sought. For example, supporters of an organization such as WWF-UK might enjoy 'consuming' the fact that they are members. The 'investment' of supporting a group can therefore be part of the individual's creation of his public 'self' – as much as the fact that he or she buys sports shoes at £90 rather than £30. To put the WWF-UK badge in the Volvo window is to advertise the membership as much as the group.

Support for a cause might well *not* be very strongly associated with the likely success of the group. An individual might enjoy the social distinctiveness of being associated with a minority cause. Of course, only a small minority of the population might adopt such a position, but as argued below some groups can be viable with only a small percentage of the potential membership. Individuals may participate in apparently fruitless group activity as it is the participation that is the reward rather than the outcome. Members may be content to see activity rather than policy success. Groups can perform a surrogate function; their activity removes the need for individual participation. For an annual subscription fee members can support an organization such as Greenpeace without becoming active themselves.[17]

Rothenberg (1992) suggests that some groups can afford to be public losers in a policy-influencing sense, because they do not seek to be 'insiders'. The fact that Greenpeace does not prevent dirty beaches may have very little bearing on membership turnover. (In fact, it could be argued that the more dirty beaches there are the greater the 'need' for Greenpeace.) In fact, as Lowery and Gray (1995: 9) point out, in the environmental sphere the competitors of environmental groups are other environmental group, *not polluters*: 'Indeed, increased numbers of polluters may actually help the environmental group survive by convincing its members that work remains undone.'

Godwin (1992: 138) suggests that direct mail can lead to political negativism as groups may be obliged to tailor their political involvement towards visible tactics and emotional issues. He cites a consumer lobbyist: 'If the press isn't going to be interested, then neither are we. We have to show our members we're doing something.' If there is a perception of impotence or neglect then competitor organizations may emerge hoping to poach members. For example, in 1994 a new group called *Breach* was formed to oppose a resumption of international whaling. Breach claimed that it represented 'disillusionment' with groups such as the WWF and Greenpeace. Breach aimed to 'put an end to this barbaric slaughter once and for all. If others choose to make deals and compromise that is their affair ... some environmental groups are preparing to agree to a compromise that would see the IWC [International Whaling Commission] allocate hypothetical quotas of whaling for each whaling nation' (*Daily Telegraph*, 22 May 1994). In marketing terms Breach was attempting to position its product by use of a *Unique Selling Proposition* (USP). It portrayed WWF and Greenpeace as ineffectual in the campaign against whaling and sought to take advantage of its position in the 'market' as the non-compromise anti-whaling group.

The decision about membership rests in part on the evaluation by the individual of the value of collective benefits. The Olson approach has been that all potential members who do not join an organization are free-riding. Our perspective is that unless one actually values the group output one does not really face the choice of support or free-riding. However, the primary source of information available is the group itself. *The view on the value of the outputs is likely to be influenced by the version of reality that the group provides.* Environmental groups highlight the serious problem for the planet caused by the destruction of the rain forest or the creation of the hole in the ozone layer. These may be both real and serious problems, but it is the one-sided summary of the technical debate that is skilfully presented by the relevant groups. Newsletters and other communications (which can be regarded as a selective incentive for the member) are more important as a means by which the group can manipulate the evaluation of issues and its role in addressing them.

Kimber (1981) makes the point that free-riding is unlikely where the member is uncertain about the supply of the collective good. He suggests that it is important to distinguish between the idea of a good being supplied, and the supply of a good being certain. He argues this is particularly the case for those who are already members. As we have seen, it is common group strategy to try and convince members that a problem is on-going, that it is an ever-present threat, and that their continued support is needed to redress the situation. In such circumstances, and with groups actively encouraging worry about the uncertainty over the supply of collective goods, the individual is pushed in favour of joining. Kimber (1981: 193) argues:

An individual outside a group that provides a collective good will not join if there is no risk of the supply of the good being discontinued – that is, if the supply is effectively certain as far as he is concerned. Once the supply of the good is threatened, however, the rational individual will join the group. The individual thus appears to follow the maxim: *if the supply of the good is uncertain, join; if it's certain, do not join* [original emphasis].

Manipulating the value of collective bads

Members are encouraged to participate by exaggeration of the *consequences* of group failure and building up the assumption that non-participation by individuals will guarantee failure of the group (see below). Crenson has argued that individuals are more likely to contribute to the prevention of a collective bad than to the securing of a collective good: collective goods are paired with collective bads, 'whose marginal costs decline as the bads become more plentiful. The other side of public sanitation, for example, is public trashiness' (Crenson, 1987: 265). He pointed out that rational individuals might bear the initial costs of providing a collective good, in the knowledge that once it was provided, say a clean street, others would contribute to prevent it being lost. People are more easily mobilized in response to threats than in response to prospects. Mitchell (1979: 120) remarks:

In a situation where individuals have a high disutility for public bads that they are unable to escape, where they have imperfect information, and where the cost of contribution is low, the act of contributing is consonant with a rational strategy of seeking to minimize the maximum regret.

All this, of course, means that groups can feed the potential member information at the same time as they are seeking to present the monetary costs of joining as being negligible. Mitchell (1979: 121) argues: 'environmental lobbies have an incentive to highlight (if not exaggerate) the threat posed by environmental bads and/or the desirability of the collective goods they seek to obtain and their organizational effectiveness in preventing bads or obtaining goods'. Bosso (1995: 114) suggests that: 'direct mailers shop for the next eco crisis to keep the money coming in'. Bosso's *Ambulance Chasing* argument is supported by Pierre Gassmann of the International Red Cross who argued that (in Germany): 'Only those who have a "sexy catastrophe" on offer ... can rely on sufficient donations from sensitive European humanists' (*SPIEGEL Special*, November 1995).

Godwin (1988: 26) shows that empirically negative assertions are important in direct mail. A content analysis of a range of solicitations found that 60–70 per cent of the paragraphs contained information, 30–40 per cent played on citizen duty, and approximately 35 per cent attempted to motivate by stressing fear or guilt. Godwin (1988: 47) said that only 5 per cent of the paragraphs invoked solidary or material incentives. The tactic of stressing collective bads can sometimes be *demotivating*. In a 1992 survey of its lapsed members Greenpeace UK discovered that some members left the group because of 'doom fatigue'. A Greenpeace spokesperson commented (in *The Observer*, 27 June 1993) that: 'They [members] complained of being helpless, that they could not do anything about environmental problems. They leave to get away from all the bad news.' It would seem on this (limited) evidence that it is important to at least some members that the group is successful and not just active in its campaigning. The article went on to note that as a result of these findings groups like Greenpeace and FoE had changed tack by switching their membership strategies to concentrate on delivering solutions rather than emphasizing problems.

Manipulating perceptions of personal efficacy

Olson suggests that individuals will make logical calculations that would recognize that outcomes are likely to be independent of whether or not they join. The planet is unlikely to be much affected by whether or not one gives to Greenpeace or spends the money on a consumer durable. However, in practice individuals seem to have exaggerated conceptions of their efficacy; they do expect their action to influence the delivery of valued goals.

In Table 5.4 we reported that 71.2 per cent of new and 74.7 per cent of veteran FoE members, and 69.6 per cent of Amnesty members believe that their support made a difference. Indeed, the most interesting data relates to the exceptionally few respondents who felt that their participation had no effect (7.3 per cent FoE veterans, 6.9 per cent of Amnesty members). These responses suggest that group members do not accept the free-riding premise that non-membership is marginal to the group.

Olson suggests that potential members will assume that their action is irrelevant to group viability and that they can safely free-ride. However, the psychological pressure to 'buy one's round' works in groups – especially when the costs are low. Dunleavy (1991), citing Finkel, Muller, and Opp (1989), notes that groups develop an ethical duty on potential members to participate and not free-ride, and the empirically widespread belief in the 'unity principle' – which holds that 'the participation of everyone (in the identity set) is necessary to have a chance of obtaining the (group's) public good' (Dunleavy, 1991: 62). Belief in the unity principle seems more widespread than an urge to free-ride (see Marwell and Ames, 1981; and Morrison, 1979). This 'equity ethic' seems stronger when others are seen as contributing. Somewhat paradoxically, however, the more visible the contribution of others, the *less likely*, in our view, is free-riding. This may not be an Olson concept of rationality, but it may again reflect the assumptions of large enough numbers to make groups viable.

Potential members apply the principle that if they free-ride, then it is reasonable to assume that others will. This may be illogical in Olson's terms but it seems a reasonable assumption in practical circumstances. Moreover, group communications encourage these beliefs. Amnesty appeals confront denial and assert: 'You can do something' (Cohen, 1995: 124). While a Greenpeace appeal to its lapsed members said:

Greenpeace only exists because people like *you* continue to be supporters.

Without *you* there is no Greenpeace. And without *your* past support actions like the ones below may well have been impossible. We can't carry on unless our supporters, people like you, renew their support ... Our actions in the future depend on you acting *now* ... *Without you we're sunk. With you we can change the world* [original emphasis].

When the role of personal efficacy is combined with our other major point – that one cannot free-ride if the act of subscribing is itself valued – then the potential for group viability is further enhanced.

The stress on marketing in this chapter attempts to explain the large increases in membership for some public interest groups despite the threat to mobilization posed by the free-rider problem. As Berry (1984: 82) argued over a decade ago: 'For some interest group leaders, then, it may not be the mix of benefits that they offer, but how aggressively they market their organizations, which makes the crucial difference.' The success of marketing has, however, led to low levels of membership stability. To regard environmental groups as protest businesses opens the door to a discussion of business viability: inter-group competition for supporters can be fierce. Imig (1994: 19) identified a process of groups attempting to *segment 'their' markets* within the 'Children's Lobby' in the United States: 'as with other public interest groups, children's advocates work to differentiate between themselves in as many ways as possible, including their programmatic foci, their sources of revenue, and the services they offer members'. Bosso (1994: 47) asks how many (large environmental organization) niches will the overall 'market' bear? He notes the stiff inter-group competition in the 1990s and predicts 'this sector of the advocacy community is due for a serious shakeout'.[18]

Reducing the costs of membership: patronage

Most groups try to deal with membership instability by supplementing subscription income with other forms of finance. The organizational strategy is to develop a financial portfolio which contains various income-generation sources. Not only will this help insulate the group from any fluctuations in subscription income, but finance from other sources can be used to subsidize the costs of membership.[19] Walker (1991: 33) pointed out that the Federal government in the United States (like governments in most democracies) is an important sponsor of many interest groups. However, while the national government 'is unlikely to be the sole sponsor of a group ... its overall impact on the group system, through tax incentives, contracts, and grants, is enormous ... Government support is an important source of patronage for groups in the United States, as elsewhere.'

Walker (1991: 48) found that bodies including corporations, foundations, government agencies, private charities, etc. 'often provide financial and organizational support that is the key to the maintenance and effectiveness of interest groups'. Thus, groups have been able to expand their financial base beyond what could be raised by membership subscriptions alone. In fact, some groups are heavily reliant on patronage for their financial survival. Reliance on patronage helps resolve the free-rider problem because individual subscriptions become less important to the organization. In fact Walker (1991: 49) argued that one of the most important reasons for the rapid expansion of the citizen groups sector in the United States was because of the growth of patronage. His 1980 survey found that 89 per cent of citizen groups had received financial assistance from an outside source in order to start its operations. In 1990–91 the Environmental Defense Fund had a total revenue of $16.9 million 'of which 22 percent came from foundation grants' (Walker, 1991: 32). While Imig (1994: 14) highlighted that in 1992 the Children's Defense Fund's total income was approximately $12.9 million, of which some 59 per cent came from private foundation grant and a further 10 per cent came from corporate donations.

In the United Kingdom many cause groups receive large sums of money from patronage (i.e. legacies, corporate donations, charitable trusts, etc.). In 1990 the RSPCA had a total income of £21.7 million, of which £17 million came from legacies and £128,000 from membership subscriptions (Garner, 1993: 46; cited in Grant 1995:32). In 1993/94 21 per cent (approximately £4 million) of WWF-UK's income came from legacies, 18 per cent from aid agencies and grant, 7 per cent from corporate donations (from organizations such as Boots, Cadbury, ICI, National Westminster Bank, Shell, and Tesco), and 2 per cent form charitable trusts (WWF-UK, 1994).[20]

Instability as a consequence of marketing: the 'revolving door' problem

These types of organization operate on a revolving door model – to compensate for high drop-out, they have to work hard to attract new recruits.

(Cohen, 1995: 177)

Many public interest groups (and political parties) face a membership turnover problem not stressed in the literature. For example, the Green Party made a net gain of more than 10,000 members between 1988 and 1990, and a net loss of more than 13,000 between 1990 and 1993 (this deficit represented a loss of half its membership).[21] Turnover rates are very high; for example, Letto (1992; cited in Bosso, 1995: 114) estimated that the average turnover rate for environmental groups in the United States was 30 per cent per annum. While Godwin (1988: 66) argued that those US groups which have in the past relied upon marketing techniques to recruit members (such as Common Cause and Zero Population Growth) have suffered 'volatile swings and permanent downturns' in their membership in recent times.[22]

With regard to environmental groups in the United Kingdom, it is important to explain why *only* 35 per cent of those who joined FoE in 1991 rejoined in 1992. Between September 1991 and October 1992 the Ramblers' Association increased its membership from 85,934 to 90,269 but this included a loss of 13.3 per cent of the 1991 membership. Samstag (1988: 149) estimated that the RSPB has an annual 'drop-out' rate of around 7 per cent: 'the Society ... has to recruit 30,000 members each year just to "stand still" '. The 'normal' Amnesty drop-out rate is 40 per cent after the first year; the average membership-life is four years (Cohen, 1995: 235). This again confirms that non-renewal is a major issue. In 1992/93, although Amnesty managed to achieve an overall net increase in membership of 16.5 per cent, it actually lost 24.5 per cent of its pre-existing membership. It had to recruit 41 per cent to attain the 16.5 per cent increase (see Table 6.2). (According to Cohen (1995: 176) Amnesty lost 20,000 members in the 1992/93 period and recruited some 29,000 new members.) Apparent stability is the net result of balancing large scale entry and exits. Drop-out rates of between 30 and 40 per cent are seen as acceptably healthy by many public interest groups.[23]

Table 6.2 *Membership turnover at Amnesty, 1992–1993*

Date	Membership
June 1992	82,919 (100%)
Joined 1992/93	33,977 (+41%)
Lapsed 1992/93	20,340 (-24.5%)
May 1993	96,556 (+16.5%)

Ironically the groups that devote most attention to recruiting and retaining membership might be particularly prone to instability. Information from FoE Scotland found that the organization was in slow growth/plateau mode from a membership of 4,169 in 1990 (1 January) to 4,487 on 1995 (1 January), but as with other public interest groups there was a turnover problem which meant that the consistency of numbers did not represent a consistency of individuals. However, the level of instability seems much lower than for FoE (London) Limited and Amnesty. In 1993 FoE Scotland had a net increase of 127 (2.9 per cent) but this masked a loss of 12.5 per cent (this contrasts with Amnesty's net increase of 16.5 per cent and its loss of 24.5 per cent. It is at least possible that the greater stability for FoE (Scotland) is the result of their lower media profile: potential members are less likely to find the group taking the initiative in giving 'joining opportunities' – members have to work (comparatively) harder at securing membership. Accordingly those who join may be more committed: aggressive recruiting succeeds (well) in stimulating those weakly predisposed to membership.

Instability is strikingly different for different groups. We assumed at first that membership turnover would be lower for non-public interest groups. This would be a partial confirmation of the Olson thesis to the extent that the incentives on offer from public interest groups may be less powerful in securing rejoining than the selective material incentives available for profit-based groups. However, Table 6.3 (tentatively) suggests that it is the organizations with the strongest advertising presence that have the highest turnover.

In Olson's perspective a rational decision on rejoining would presumably arrive at the same conclusion as that on joining, if there was not a major change in circumstances. Would Olson-type members fail to respond to several invitations to rejoin but do so at the fifth

Table 6.3 *Membership stability of selected groups, 1991–1992*

	1991	1992	Lapsed memberships	Membership retention rate[a] (%)
Profit organizations[b]				
British Agrochemical Association	38	37	1 97.4	
British Meat Manufacturers' Association	98	100	0	100.0
British Merchant Banking and Security Houses Association	51	56	1	98.0
British Retail Consortium	170	180	0	100.0
Metal Finishing Association	230	260	25	89.1
Metal Packaging Manufacturers Association	43	42	2	95.3
National Association of Tile Distributors	53	55	0	100.0
National Federation of Fishermen's Organizations	4,000	4,500	200	95.0
Tenant Farmers' Association Limited	3,650	3,100	200	94.5
United Kingdom Petroleum Industry Association	13	14	0	100.0
Citizen groups[b]				
Amnesty International	82,919	96,556	20,340	75.5
British Motorcyclists Federation	97,000	108,000	2,100	97.8
Campaign for Real Ale	30,000	36,000	2,000	93.3
International League for the Protection of Horses	44,000	50,000	4,000	90.9
Ramblers Association	85,934	90,269	11,203	87.0
Royal Navy Bird Watching Society	338	329	19	94.5
Society for the Protection of Unborn Children	40,000	47,000	1,000	97.5

Notes:
[a] i.e. % of members retained from 1991 to 1992.
[b] Terms as used by Walker (1991).

offer? In their schema would the decision on maximizing personal welfare not be the same on different occasions? Not only is the fact that members 'rationally' decide to join in response to one joining opportunity rather than others is not immediately obvious in the

Olson account, nor is the volatility of membership. As Rothenberg (1992: 20) points out, if it is argued that membership (in the face of the lack of selective material incentives) stems from an exaggerated sense of the potential member's sense of political efficacy: 'The assumption seems to be that once these contributors overestimate their impact, they repeat the same mistake again and again ... (consequently) organizational membership ought to be quite stable.'

Rothenberg (1992: 22) addresses the issue of membership instability through his theory of 'experiential search'. He argues that initial decisions to join groups are rarely rational in the way Olson predicts, but nonetheless are rational in the sense of displaying a deliberate learning process. At the point of rejoining: 'the decision to join makes sense as a strategy by individuals who recognize their lack of knowledge. Members join groups to learn about them, and as they acquire knowledge, some can be expected to leave.'

Rothenberg's work is important in taking us beyond the fixation with the initial decision to join to a focus on the decision to rejoin and remain in membership. As Dunleavy (1991: 54) has noted, public choice accounts tends to perceive 'group joining as a one-off problem'. In fact, the rejoining decision often appears to 'undo' the initial decision to join. There is high membership turnover among new members for many public interest groups.

Rothenberg interprets membership as about 'testing before final purchase'. Our main thesis suggests that high turnover is also consistent with the fact that mail order based groups permit a form of political participation which can be labelled *cheap participation*. For a cost below the threshold of serious analysis by the relatively affluent potential member, they can make a political statement of preference, without engaging in the costs (time and money) of 'real' participation. This is to say that for mail order groups *the participation demands are in fact very low*. It is the casual nature of the engagement rather than subsequent disillusionment that accounts for turnover.[24]

Building the group into the membership decision

The importance of group activity is that it places the decision on membership on to the agenda. It encourages potential members to consider the issue. Moreover, the group can actively shape the parameters of choice. It can persuade potential members that the issue is of importance, and it can manipulate downwards to cost of contribution via cross-subsidization from patronage or the provision of selective incentives. Some groups are prepared to lose money in a direct sense that direct mail does not bring in sufficient members to pay for itself but there is the educative benefit of alerting and informing those parts of the population who do not join.

The significance of individuals in group behaviour has been seriously eroded since Walker's (1991) revelations of the extent of group patronage. This is not to say that the individual motivation to join is not important, but in our view the membership of groups such as Amnesty and FoE is a reflection not just of a spontaneous public interest but of:

1 a level of concern about an issue that the groups can generate by providing information;
2 a level of predisposition to support group goals that groups can alter by their persuasion and presentation; and
3 a level of membership that groups can alter by their recruitment strategies.

Over time there has been a rejection of the unrealistic nature of Olson's assumption that: '*Rational, self interested individuals will not act to achieve their common or group interests*' (Olson, 1971 edn: 2, original emphasis). Expressive and other non-material benefits also act as selective incentives. The contribution of this literature is illustrated below in Table 6.4.

Table 6.4 *Including group initiative in the membership decision*

1.	*Inclination = mobilization*
2.	Olson suggested:
	inclination + economic calculation = mobilization
3.	The acceptance of soft incentives means that the equation becomes:
	inclination + economic and non-economic calculation = mobilization
4.	If one accepts the primacy of the group in shaping preferences and that individuals come to different decisions when given different joining opportunities, the pattern is more like:
	inclination + joining opportunities by the group + repeated economic and non-economic calculation (after group persuasion) = mobilization

Such a version of events is more realistic than Olson's in building in the importance of the group, the attraction of non-material incentives, and the reiterated nature of choice. However, we again argue that the actual decision may be simpler than both the Olson version and this revised 'test' suggests. Members *need* to make elaborate calculations about group viability only if they are giving on the basis that they will only back *successful* groups. If they pursue an expressive rather than an instrumental incentive then the decision is much simpler. 'Will I feel better about myself if I give to the RSPB? If I can easily afford it, I'll join.' The decision is not about future events for which information will need to be gathered, but an instant decision about the impact of joining in terms of self-gratification. This strategy short circuits more complex decision scenarios.

Notes

1   We argue similarly in, 'Manipulating Membership: Supply Side Influences on Group Size' forthcoming in the *British Journal of Political Science*.

2   In this chapter we use the notions of membership and supportership as interchangeable. In other discussions these terms need to be disentangled.

3   See also Johnson (1995: 10) who argues that, 'the recruiting efforts of environmental groups are a critical part of the explanation of group membership patterns ... membership patterns depend vitally on the recruitment activities of groups'.

4   This is similar to Putnam's concept of 'tertiary associations' discussed in chapter 7.

5   It has been persuasively pointed out to us by an anonymous reviewer that this point is a confirmation of Olson's approach: groups have deliberately designed themselves that way.

6   But it has now also started to market selective incentives.

7   The response rate from a mailing is substantially affected by the quality of the mailing list. One technique used by groups is to 'piggy back' on a general consumer survey as to whether or not members of the public would consider supporting the cause in question: a mailing to those who have defined themselves as likely to give support is obviously much more successful than even to a list of supporters of similar causes.

8   The marketing argument can be further extended to the lobbying strategies of protest business. For example, Steve Sawyer, former Executive Director of Greenpeace and former head of Greenpeace USA has said that: 'The nature of the central ideological debate within Greenpeace is: do we

approach direct action as a political tool or as a marketing trademark? ...
The moral and political quality of what we do is what distinguishes us from
a stunt-making, money-making machine' (*Telegraph Magazine*, 11 May
1996). Whether or not one accepts Sawyer's justification of the motivation
behind Greenpeace's tactics, it is clear that its tactics have established its
name as a valuable trademark. As John May points out, Greenpeace is 'one
of the most valuable trademarks on the planet, its name recognition up
there with Coca-Cola' (*Ibid.*).

9   Johnson (1995: 28) is quoted extensively as his concerns closely
match those of this book and his data is rich. He quotes one respondent
explaining a drop in membership in the early 1990s as follows. 'We had
mismanagement in the membership department, it was internal problems
... [T]hey rolled out with a direct mail campaign, a new control, without
testing it, and it was not too successful, and then they used it again in
another campaign ... They stopped doing renewal telemarketing and they
cut the last two efforts of the renewal series plus two renewal wraps on the
magazines that were sent out.' In other words the explanation of the drop
in membership was explained in terms of group skill rather than changes in
the potential membership in terms of attitude or lack of funds in the recession. We concur with Johnson's (1995: 28) conclusion that a widespread
collective interest does not itself generate a group: 'Rather, to understand
the groups that exist, one must understand that there are organizers and
recruiting strategies whose efforts must somehow interact with the attitudinal basis of support in the community.'

10   Johnson (1995: 17) conceded that not all groups were sufficiently
professional to have followed this 'good practice'. He also (1995: 17)
described the 'control' mail package which groups have developed as their
most successful option and which they normally use, but they also regularly
test this against the response rate to different versions. One respondent told
him: 'So what becomes my "continuation", or control" package, is the
product of head-to head testing ... Everything is done on the basis of cost
benefit analysis.'

11   Dowding (1991) has suggested that one response by groups to
potential free-riding is the supply-side attempt to convince would-be free-
riders that if they do not contribute then they will not receive the good
because it will not be provided at all without their 'crucial' contribution.

12   The expectation in public choice accounts (such as Olson) appears
to be that group membership is driven by inducements such as discounted
hotel accommodation, cheap insurance, private health care and so forth.
These sorts of considerations are particularly associated with sectional
groups such as unions or trade bodies but public interest groups also offer
selective benefits. The WWF offers various incentives, ranging from site
visits to newsletters. The ETA gives certain benefits such as a help-line, legal

expenses for accidents involving personal injury while travelling, and car rental discounts. These selective benefits may seem trivial, but at low subscription levels they can be significant. If we extend our marketing theme the 'free gifts' that are on offer are not simply catering to a materialistic, selective benefit side of potential members but, from a supply-side perspective, they are 'loss leaders'.

13   Some groups who cannot afford the high costs of advertising have recruited support from 'high-profile' individuals who are able to attract attention to the groups' cause. Compassion in World Farming (CIWF) has recruited Joanna Lumley, Penelope Keith and Julie Christie. Mrs Joyce D'Silva, Director of CIWF said that their campaign against the exportation of calves from the United Kingdom in 1994 'really began a year ago with Joanna Lumley crying at a press conference when we showed a film of calves in veal crates ... The photographers kept clicking away and the pictures appeared everywhere ... membership soared' (*Daily Telegraph*, 19 January 1995).

14   We are grateful to Dr Klaus Schubert for drawing our attention to the parallels between our argument and the current ongoing debate in Germany.

15   As Johnson (1995: 19) explains: 'If a group mails 10,000 pieces at $.35 each, at a 1 per cent rate the organization spends $35 dollars to recruit each individual member. The member pays $25 in the first year, so that the group not only loses money on the acquisition but also ends up providing the benefits of the group free during the first year. But the group expects between 30 and 50 per cent of first-year members to renew, and the renewal rate climbs near 100 per cent over the following three years. "So what you are doing is looking at the cumulative value over time of those hundred members".'

16   The Royal National Lifeboat Institution (RNLI) has five categories of membership: Shoreline membership (minimum £10 per annum); Joint Shoreline membership (minimum £17 per annum); Governorship (minimum £33 per annum); Life Governorship (minimum £500 one-off payment); and Storm Force membership (under 16s minimum £3 per annum).

17   Again we are aware of the dangers of generalizing across the range of interest groups. Some groups may be engaged in activity where their members perceive the costs of policy failure far more highly. Our working assumption is that the potential member of most large public interest groups is looking for evidence that the group will attempt to address the relevant problem, rather than a guarantee of success.

18   Bosso (1994: 47) notes that: 'a Clinton administration more hospitable to the environmental agenda ironically exacerbates the problem because scaring contributors into writing checks is going to be a lot harder now that the vice president is an environmental leader'.

19   For example, an RSPCA official we interviewed claimed that the group was previously subsidizing its membership via its income from other sources.

20   In fact, not only is patronage important/crucial for group maintenance, it is also very important in the early stages of group formation. According to Nownes and Neeley (1996: 124) patronage allows many public interest groups not so much to *overcome* the "free-rider" problem as *by-pass* it. They argue that both Walker (1991) and Hansen (1985) see mobilization as more dependent on patronage than members: 'The key to understanding which organizations form, then, is understanding which groups get subsidized and when' (Hansen, 1985: 94).

21   Rüdig *et al*. (1993: 45) carried out a panel survey of Green Party members and found that: 'Of all the members surveyed in November 1990, 30% had joined in 1989 ... By November 1991, 40% of 1989 members who responded to our second questionnaire had ceased to be members or were about to leave the Party. And by November 1992, almost 50% had left the Party.'

22   Indeed, even though Godwin's direct-marketing recruits thought a group was successful (this may be explained by their reliance on group leaders for information) their membership was still likely to be less permanent. Godwin (1988: 65–6) observed these patterns may have serious implications for citizen action groups dependent on direct marketing for members and income. These groups may suffer substantially greater losses than social network associations when an issue loses its position in the political limelight.

23   Mundo (1992: 177) reported that in the US 'the Sierra Club enjoys a healthy [*sic*] 73 percent renewal rate'.

24   This is why groups encourage members to pay their annual subscription via direct debit. This method of payment removes the membership renewal decision from an individual's immediate agenda.

7

Protest businesses and democratic activity

Where few take part in decisions there is little democracy; the more participation there is in decisions, the more democracy there is.

(Verba and Nie, 1972: 1)

Is 'more', more influential?

Two importantly different statistics are relevant in trying to gauge the impact of public interest groups. The first relates to the absolute number of organizations, while the second looks at the rate and level of mobilization achieved by the groups. On the whole there is an assumption that numbers of groups have increased (i.e. what is termed 'density' by Lowery and Gray (1992)), but the data to support that assumption is not overwhelming. That conclusion is usually a response to the pressure of the newspaper clippings that cite new name after new name rather than rigorous quantification. In noting the splintering of some long-standing groups (e.g. the creation of FoE by disaffected Sierra Club activists) and the emergence of 'clean sheet' organizations, we tend to *assume* that the overall population of groups, and the mobilization of individuals within groups has to be growing. However, there is a less well noted mortality among groups: the fact that new bodies emerge does not necessarily mean an overall increase.

So though the popular (and not necessarily wrong) assumption in this sub-field is that there has been a year-on-year increase in the number of organizations and the level of their support, Tables 1.1. and 1.2 in chapter 1 suggest that, at least as far as the environmental sphere is concerned, membership has reached a plateau, if not beginning to fall.

An increase in activity (however measured) encourages some to see the political process as more pluralist and hence normatively acceptable. However, the contribution by Lowery and Gray (1992) is that the *nature* of the newly mobilized groups needs to be taken into account: if *more* simply means *more business groups* then the system is not fundamentally altered by the increase. There can be increased *density*, without usefully increased diversity. Therefore the optimistic interpretation – that the new environmental groups are effectively changing political outcomes needs to be viewed with a degree of scepticism:

1   Is the growth in the environmental population 'net'? Is there a mortality among earlier groups?
2   Is the public interest growth matched or exceeded by business and other non-public interest groups?
3   Even if there is a large number of public interest groups are they nonetheless strategically disadvantaged in terms of access and political influence?

There need be no immediate and automatic connection between 'more', and 'more influential'.

The decay of parties and the growth of groups?

It has been claimed that the expansion of the group system is causally linked to the weakening of political parties.[1] The traditional view is that parties were the best representative agency available. Schlozman and Tierney (1986: 201) say that: 'party and organized interest strength are proportional: where parties are strong organized interests are weak and vice versa'. As Peterson (1992: 238) has pointed out, Walter Dean Burnham (1970: 133) argued: 'political parties ... are the only devices thus far invented by the wit of Western man which with some effectiveness can generate countervailing collective power on behalf of the many individually powerless against the relatively few who are individually – or organizationally – powerful'. Peterson goes on to quote Schattschneider's (1942: 192) claim that parties and groups were in a zero sum competition: 'If the parties exercised the power to govern effectively, *they would shut out the pressure groups*'. Dahl too is cited as repeating the idea that: 'the strength of parties in policy-making tends to be inversely related to the strength of pressure groups' (Peterson, 1992: 238).

In fact, parties have perhaps always been less successful than the theoretical role ascribed to them. Almond and Verba (1963: 192) in their discussion of their major cross-national study of participation argued that: *'Clearly, no matter how important the role of political parties may be in democratic societies, relatively few citizens think of them as the first place where support may be enlisted for attempts to influence government'* (cited in Richardson, 1995: 118, emphasis added). It is now increasingly being argued that participation, once seen as primarily pursued through political parties, has been replaced by group membership.

The party/electoral channel of representation has been increasingly seen as deficient. The breadth of support on which successful parties rely is seen, in one sense at least, as a democratic weakness: the party often cannot afford to reflect the narrow, and intensely held, concerns of individuals (see Cigler and Loomis, 1995 edn: 19). These, it is argued, can be better expressed through single-issue groups. Berry (1984: 55) suggests that if one is politically active and willing to spend $100 to pursue one's political goals in a year then the interest group can be seen as a better 'investment' in that one can better target the cash on one's personal priorities be they protecting wilderness or whatever. Elsewhere he argues that: 'The essence of the public interest philosophy is that party politics cannot be trusted. Parties are impure; they stand for compromise rather than hard-fought principles' (Berry, 1980: 43; quoted in Hershey, 1993: 148). In this view, the increase in the number and size of campaigning groups (particularly environmental groups) has shifted the channels of politics (from parties to groups), increased participatory democracy through organizations which more clearly articulate views than do broad church parties, and suggests a turning away from the values that are represented by traditional parties. Loomis and Cigler (1991: 19) have argued that the traditional party system found it difficult to deal effectively with citizens' high expectations and a changing class structure.

It is within groups rather than parties that much recent participation has been identified. Katz argued that there has been the erosion of sociological ties between individuals and parties in Europe (quoted in Richardson, 1995: 119). Seyd and Whiteley (1992: 204) have reported the view that: 'alternative forms of participation include single issue pressure groups, and new social movements ... provide a more rewarding type of political participation for many people than membership of a political party'. Lawson and Merkl (1988: 3) speculate that: 'It may be that the institution of party is gradually disappearing, slowly being replaced by new political structures more suitable for the economic and technological realities of twenty-first-century politics.' Accordingly, *alternative organizations* are emerging as *would-be surrogates* for parties (Lawson and Merkl, 1988: 5). Environmental organizations are one of the four new types of emerging alternative organizations referred to by Lawson and Merkl.

Between 1952 and 1991 Labour Party membership declined from 1 million to 320,000 (Butler and Kavanagh, 1992: 59) and to 279,530 in 1992 (Webb, 1994: 113; Ware 1996: 106).[2] Butler and Kavanagh (1992: 59) argued that even the 320,000 figure 'greatly exaggerated the actual number of paid-up members'.[3] A similar picture emerges with regard to the Conservative Party. Webb (1994: 113) estimated that between 1953 and 1992 Conservative Party membership fell from 2.8 million to 500,000.[4] Lipow and Seyd (1996: 276) warn that the data on party membership levels needs to be viewed with a degree of scepticism. Membership figures were 'compiled through the voluntary efforts of political amateurs in party branches whose reliability cannot be guaranteed'. However, they concede that the caveats they raise do not 'question the decline of party membership, but ... suggest that the decline may not be as precipitous as the raw figures suggest'.

Richardson (1995: 121) in trying to assess the decline of party organizations in Western Europe concurred with Gallagher *et al.* (1992) who argued that while 'no clear trend emerges ... the weight of the evidence suggests that the decline theory is correct'. Ware (1996: 74) cites Katz and Mair's (1992) data as confirming 'the widely held belief that party membership and more generally activity within parties has tended to decline'.

In contrast to party organizations, interest groups have experienced dramatic growth rates. For example (as Table 1.1 illustrates) FoE has grown from 1,000 members in 1971 to around 120,000 in 1993. In 1993 Greenpeace had 410,000 members and the RSPB had 850,000 (CSO, 1995: 189).[5] Writing in Britain, Jonathon Porritt (*Daily Telegraph*, 7 October 1995) referred to the numbers of paid-up members of environmental organizations as 'a vast green army on the move, massively outweighing the collective membership of all political parties combined'.

Thus a direct relationship between the decline in the membership of political parties and the rise in non-party political organizations has been suggested: one behaviour apparently replacing the other (see Grant, 1995: 1, 81). As a consequence of this increased group mobilization Dalton (1993: 8) argues that citizen groups are 'transforming the nature of contemporary democratic politics'. Arguably these groups have increased the opportunities for, and involvement in, participatory democracy. In 1996 in Britain a coalition of cause groups (*Real World*) was claiming that it had over 2 million supporters, apparently in excess of the collective membership of political parties.

Related to the discussion of the decline of parties, then, is an implicit argument about the superior type of participation 'new politics' organizations offer their 'members'. Groups, it has been argued in general terms, offer a more satisfactory sort of participation than available through the party and electoral channels. As Richardson (1995: 128–9) highlights: 'Environmental organizations are seen as a response to today's "New Politics" ... [and] have a different participatory style within their organizations, said to be more in tune with the aspirations of the activists who join.' Belief in the efficacy of party participation eroded as parties became very broad office-winning machines that tended to avoid making specific programme commitments. If an individual backed policy change on issues such as abortion, pollution control, or logging, they were likely to gain more satisfaction in a single issue group than a compromise party. Within a group the participation seemed more immediately connected to a policy position. The individual's effort/contribution is allocated to a specific purpose.

Two discrete modes of participation have been identified – *elite-challenging* and *elite-directed* (Inglehart and Klingemann, 1979). *Elite-challenging* political participation, it is argued, is synonymous with 'new politics' organizations and entails the direct involvement of the individual '... in specific political decisions (i.e. direct democracy) and can be explained by the growing disposition to use unconventional, sometimes even illegal, forms of political action in order to influence political decision in the desired way' (Poguntke, 1993: 136).

Apart from what we might call the targeting of participatory effort in groups the membership experience is argued as being more democratic. *Elite-directed* political participation is essentially 'reactive

political behaviour. The individuals choose between alternative political packages that are usually presented by political élites' (Poguntke,
1993: 136). It is *claimed* that bodies such as the Green Party, Greenpeace, FoE, and Amnesty exemplify *elite-challenging* behavioural
patterns and the concept of *Basisdemokratie* (i.e. grassroots or direct
democracy), while in 'traditional' political parties members are more
predisposed to *elite-directed* behavioural patterns.

However, while party membership totals may be in decline, participation in large-scale interest group politics need not be extending participatory/direct democracy in a compensatory way. Though
in terms of broad quantification group activity may have surpassed
party participation, it does not necessarily imply that what is
involved is more than a minimalistic, relatively small, financial contribution. The 'members' of groups such as FoE, Greenpeace and
many other direct-mail organizations have no more (and most probably even less) control of their organizations than do party members.

'Membership' of large-scale environmental groups may be significantly different from the sort of direct action participation discussed by Dalton (1994) or Poguntke (1993). Commenting on the
work of Bosso, Cigler and Loomis (1995: 101) have observed a
change in the US environmental movement in the past three decades
from:

a handful of groups staffed by largely self-sufficient and part-time cadres
of amateurs dedicated to wildlife protection and land conservation, the
movement has evolved into a diverse collection of highly professional
organizations ... Environmental organizations involved in contemporary
Washington politics are characterised by their extensive technical and
scientific expertise, deep legal talent, professional management, state-of-
the-communications and fundraising technologies, and sophisticated educational and public relations campaigns, as well as by their use of direct
lobbying.[6]

The conflict between success and influence?

Groups wishing to influence the shape and direction of public policy
effectively have two main routes to achieve their objective: i.e.
inside or outside channels – with the inside channel generally seen
as being the more likely to be effective. Grant (1978), and Schlozman and Tierney (1986: 169) maintain that groups have a choice

over which strategies to pursue: 'Once strategic matters have been settled (the institutional arena in which political conflict will take place), an organized interest still must choose among assorted political tactics.'

However, the choice of strategy is not solely related to affecting outcomes, it can have a significant impact on organizational maintenance. Following Wilson (1973: 30–2) Walker (1991: 104) argued that the policy influencing strategy has to be compatible with the need for organizational maintenance: 'The first priority for groups leaders is to find an organizational strategy that will ensure the continued existence of their group, and their choice of tactics usually arises out of this search.' Walker (1991: 105) further argues that: 'The character of a group's membership is another determining factor.'[7] Organizational survival is dependent on pursuing those values and policies which are of direct importance to group members. Citizen groups must continually reinforce members' loyalties: 'Much of what they know about the achievements of their group they learn through reports in the mass media, so that citizen groups are almost forced into an outside strategy of public persuasion and political mobilization' (Walker, 1991: 105–6).

In his 1988 study of the Campaign for Nuclear Disarmament (CND), Byrne argued the organization was 'forced' to pursue its objectives via 'unconventional tactics' because of the political opportunity structure it faced. Nevertheless, he conceded that:

Even if the political opportunity structure were such as to offer CND the possibility of using only conventional tactics, there is a substantial body of opinion among the membership who would view this as inadequate ... were CND's leadership actively to discourage unconventional protest, they would risk alienating many of their most energetic supporters. (Byrne, 1988: 132–3)

Dalton (1994: 10–15) also sees political opportunity structures as a function of the group's ideological position. Thus, he argues, that while the WWF may successfully collect business donations, this option is not open to Robinwood, Earth First! and other radical environmental groups. He says that just as it would be out of character for the Civic Trust to mount a spectacular mass protest to increase funding for historic sites in Britain it is difficult to imagine the anti-system Robinwood engaged in neo-corporatist negotiations in Germany.

While groups such as FoE have found it useful to pursue their aims via numerous strategies and tactics and not simply protest, there are significant implications for groups who believe that they are likely to be 'more effective' via inside channels: a change towards more 'insider' policy influence by environmental groups might have consequences for membership. Stanfield (1985: 1350) argued that in the United States a change to responsible/insider policy-shaping was the main feature in the field of the 1980s. He cited the Executive Vice President of the National Wildlife Federation (4.5 million members):

Moderate voices that try to achieve consensus on very, very difficult issues in the long haul are going to be more effective than kamikaze environmentalists. Any of us can firebomb into an issue and perhaps stop the issue, but, in the process, blow ourselves away. But that's not being effective, and that's not winning in the long haul.

The 'insider' conclusion was also reached in Britain. The Conservation Director of Plantlife dubbed the environmental movement as the 'alternative civil service' (quoted in Rawcliffe, 1992: 5). In line with the development of more conventional and bureaucratic styles of organization, noted in chapter 1, there has been an associated change in campaigning style. Rawcliffe (1992: 6) reported that:

The 1980s saw a changing balance between the conflict and consensual approaches to campaigning as issues became both more established (or mainstream) and increasingly complex ... the greater receptiveness of government and business to green issues and the more open regulatory environment which materialised, has also had a profound effect on the development of the tactics employed by the groups ... In general, the 1980s has seen the increasing institutionalisation of the environmental groups, marked by their gradual acceptance by, and sometimes within, policy communities. Influencing policy by seeking accommodation, through consultation and bargaining rather than protest, is now a real option open to the groups which over the period have developed an extensive knowledge of those sectors of government that affect them ...

In a generally appreciative obituary for Andrew Lees, campaign director at FoE, in January 1995, Jonathon Porritt nonetheless suggested that the campaigning approach of Lees was perhaps outdated. Porritt said that Lees had believed that to be a proper friend of the earth required a constant 'outing' of its enemies. Porritt argued that such confrontational tactics were perhaps unproductive: 'What is the validity of a moral universe that recognises only the black and the white (?) ... is unyielding confrontation the best route to finding the solutions on which we all depend?' Porritt's attitude resembles that of the US environmentalists in the period of good access of the Carter administration: 'Before, we filed lawsuits and held press conferences. Now we have lunch with the assistant secretary to discuss a program' (quoted in Lester, 1995: 129).

Direct mail organizations such as Greenpeace (FoE and Amnesty) have to walk a precarious 'membership tightrope'. On the one hand, many Greenpeace members have ideological reservations about too close an identification with business conjoined with the fact that independence from business is seen as a plus. On the other hand, these organizations' membership have a 'need' for some tangible policy success which may involve consultation and/or negotiation with business interests. As Bosso (1991: 165) has pointed out: 'Even the most dedicated of checkwriters must be convinced continually that their contributions are having real impacts, and, given the heightened competition for the environmental dollar, fund-raisers face constant pressures to stir the public and keep money flowing.'[8]

The tide has not, however, fully turned towards the insider style. Many campaigning organizations have found the insider world restrictive – there are costs involved in becoming an insider. The most significant cost is the acceptance that change is more likely to be of an incremental manner. As we have argued elsewhere:

Within policy community style politics the process of policy change can be characterised as bargainable incrementalism. Acceptance of this principal rule – premised upon the shared attitudes and values of community members – shapes participants' behaviour. If bargainable incrementalism is not acceptable then groups must find another mode through which they pursue their goals. A group which rejects bargainable incrementalism excludes itself from that policy-making arena. (Maloney, Jordan, and McLaughlin, 1994: 36–7)

As a consequence of the need for acceptance of the limitations of the inside track it may already be being questioned. The Pedestrians' Association in the United Kingdom was the archetypal responsible interest group, but by 1995:

after spending 66 years talking to government ministers and local authorities members have decided that change is not coming quickly enough. In

April they recruited some M11 (anti-road building) protesters to join their own direct action, and together young and old descended on a north London street to 'bounce' illegally parked cars off pavements. (*Sunday Times*, 30 July 1995)

Bosso claims that despite conspicuous successes such as the 'veto' of a bill to open the Arctic National Wildlife Refuge to oil exploration, some mainstream environmental groups such as the Sierra Club have become frustrated by insider politics, and in the 1990s 'the emphasis appears to be shifting away from working in government to working on government' (Bosso, 1994: 40). He further argues that: 'the results have been mixed at best, and leaders of the mainstream groups have begun to more openly question the utility of traditional lobbying activities'.

Bosso (1991: 165) (correctly) concluded that citizen organizations tend to be schizophrenic because of the inherent contradictions for such groups in attempting to play a *credible 'Washington game'*. Major environmental organizations play a low-profile *insider game* arguing their technically sophisticated cases with policy-makers. However, the (iron) law of organizational survival requires these groups simultaneously to demonstrate to members that they are:

unswerving defenders of core values ... touting 'moderate' tactics and 'successful compromises' hardly is the stuff that stirs the blood of the true believers, who, after all, have joined out of their faith in the cause. The result for the organizations themselves can be a bit schizophrenic: Dr. Jekyll in government affairs negotiates directly with industry lobbyists even as Mr. Hyde in fund-raising paints these same industries in the darkest of terms and warns gloomily of the Apocalypse should members not pay their dues. This dynamic is rather common within the broader lobbying community – all interest groups play pretty much the same game – which probably is of little solace to critics who see the national groups as so constrained by organizational maintenance pressures that they have lost their lead in espousing environmental values. (Bosso, 1991: 165)

The objectives of organizational maintenance and survival provide strict disciplines on all organizations. Not only can they 'dictate' the choice of lobbying strategies and tactics, it also leads groups to develop the search for a specific 'issue niche' over which they can (hopefully) have monopolistic occupancy and control (Browne, 1991).[9] Rawcliffe (1992: 5–6) argues that 'new' environmental organizations 'have tended to colonise new "niches" within the environmental movement such as the provision of specific services or on more traditional lines, the mounting of campaigns for parts of the biosphere previously uncatered for'. While in the United States, Bosso (1991: 164) pointed out that mainstream organizations consciously occupy more or less well-defined niches in the overall policy agenda: 'the Sierra Club leads the fight against the oil industry over pollution of coastal waters ... and the National Audubon Society maintains its long primacy over wildlife issues'.

Such claims could be discounted as minority opinion based on their organizational needs – except that there is a theoretical explanation as to why environmental groups might have reservations about the insider role. Walker's argument was that internal organizational demands reduce the area of choice about strategy's relationship to the wider political system (see also Godwin, 1992). Greenpeace, for example, has a policy of not negotiating with government or business, combined with its high-profile public 'stunts', aimed at satisfying its membership, as much as achieving any policy success. It has been argued elsewhere that the need to satisfy supporter prejudice has driven Greenpeace's selection of political priorities. The *Daily Telegraph* (12 June 1995) claimed that Greenpeace's silence on fishing was because it was feared that its income in Holland, Germany, and Denmark might go down as a result. The newspaper concluded that: '... the primary objective of a modern pressure group is its own perpetuation'. The *Sunday Telegraph* pursued the same theme (10 September 1995). Matt Ridley claimed that Greenpeace:

is an organization devoted to growing as an organization; it spends most of its money not on the environment but on stunts, billboards, ships and salaries; self-promotion ... So, if Greenpeace is not a scientific or a political or an environmental organization, what is it? The answer clearly is that it is a commercial concern. It raises money from its members and donors with the intention of spending £90 million a year on things that will attract more members and donors while paying the salaries of its employees ...

According to the *Independent*, the Brent Spar campaign (to stop the dumping of a disused oil installation) was born in a suggestion from a Greenpeace employee to mount a high profile stunt to reverse the decline in membership of the previous few years.[10]

In this perspective these organizations effectively have two intertwined goals: the securing of funds from a geographically disparate membership to ensure organizational survival, and the pursuit of goals which meet the organizational *raison d'être* and consequently maintain the flow of membership funds. These two cannot be divorced. Failure in either area may lead to organizational failure and disbandment. Greenpeace is to a large extent *forced* into pursuing an outsider strategy because the perception of effort (whether it be seen as successful or effective, or just keeping an issue alive) is *all* in terms of member motivation. For the idea or cause group, the imperative of members' satisfaction inevitably leads to outsider type campaigns aimed at 'public persuasion and political mobilization'. And, of course, the continual reinforcement of the existing memberships' commitment to the cause.

So it can be argued that organizations which rely on direct mail are trapped into a high-profile strategy and 'ambulance chasing environmentalism' that seeks to dramatize and exploit issues (Rawcliffe, 1992: 9). This may not conform to the effective insider style of influence but it may be the sort of campaigning that wins financial support.[11] By 1996 Greenpeace was moving its strategy to 'solutions campaigning'. This might lead to increased effectiveness in terms of policy-change, but be a less effective means of maintaining membership.

Customers or members: democratic participation?

The idea that citizens are involved in 'politics' in addition to voting is seen as normatively desirable by many commentators: it is seen as a challenge to a sort of controlled and constrained party participation. Goldsteen and Schorr (quoted in Lester, 1995: 27) see the development of direct action as involving a turning away from representative democracy to participatory democracy. As discussed above, the decline of the party thesis and the direct replacement with groups is seen by some commentators as increasing the direct participatory element in European and North American democracies. However, Corsino (1986: 145) following Habermas (1976) (and, implicitly, Michels, 1919) sees *civil privatism* as the dominant form of political participation for the vast majority of citizens. A key component of this is the electoral system which encourages individuals to undertake minimalistic involvement via restricting political activity to voting. Political elites are held to benefit from this form of (effectively passive) political participation because 'the state' can claim legitimacy through the extent of involvement in the electoral process, while the vast majority of citizens still have no real influence.

Corsino (1986: 146) argues that elites(/political entrepreneurs) attempt:

to manipulate participation so that citizens participate loyally but have relative little substantive effect on the 'process of political will formation' (Habermas, 1976). Elites must structure campaign participation so that the majority of volunteers are 'active, yet passive; involved, yet not too involved; influential, yet deferential'.

He sees this manipulative process as breeding 'political schizophrenia'. In Corsino's (1986: 146) terms 'élites have a greater opportunity to rule and make authoritative decisions but still perpetuate the belief that the citizen has a voice in his or her destiny'. For example, even though the Labour Party's constitution states that the party conference is sovereign and takes great pride in it democratic form: 'the parliamentary leadership has shown that it is quite prepared to overlook parts of the conference-made programmes in drafting election manifestos' (Webb, 1994: 119).[12]

Although Corsino's discussion largely relates to participation in the electoral process, this appears analogous to the situation in 'memberless' groups such as FoE and Greenpeace. While looking like 'new politics' groups advancing direct democracy through the mobilization of greater citizen involvement in the political process, we would argue that *de facto* there is little in participatory terms which distinguishes them from political parties. Political parties, according to Webb, expect very little in terms of 'active' participation from their members: 'typically, prospective members are obliged to declare a broad allegiance to the stated principles of the party and to make some kind of modest financial contribution' (Webb, 1994: 112).[13] Parties attempt to mobilize as big a constituency as possible for reasons of legitimacy, and organizational maintenance and survival, but they seek followers not policy-making members. The Labour Party leadership in the United Kingdom by the mid-1980s believed that the profile of the activist element of the membership was detrimental to the party's electoral fortunes. The activists were viewed as unrepresentative of Labour's electorate and a handicap in the electoral contest. In 1995 the heavy press advertising of the Labour Party was making (successful) recruiting noises that exactly fit the 'quiescent supporter' style:

It only takes a few minutes to join the Labour Party; there's a vast untapped potential within our nation, waiting to be brought to life by a government with real vision. It's a vision that you can help to bring closer to reality ... (after) the next election ... you'll know that you helped make it happen.

The invitation is to contribute money: there is no policy-making invitation or expectation that membership involves active campaigning. The message is that membership is about giving the party the money to do the job. The leadership of FoE and Greenpeace want the same kind of vicarious 'commitment' from their 'supporters', they want to limit their participation to sending in the cash to support campaigns selected by the organization – *supporters should be seen and not heard*.

Campaign organizations have become bureaucratized and hierarchically controlled. The elite or policy entrepreneur controls the policy agenda while the volunteers do the 'depoliticized' mundane work of sending in the funds, selling raffle tickets, or buying goods from catalogues – and there is as much of a danger in glamorizing this as describing it as being meaningfully involved in the political process. Ware (1986) has argued that involvement in institutionalized politics inexorably draws parties which are committed to *Basisdemokratie* (i.e. grassroots or direct democracy) away from such principles: 'Even those parties (such as socialist ones) which were supposedly committed to some form of control from the base found that the institutional structures they faced were ill-suited to this, and provided little opportunity for the expansion of party activity at the mass level of politics' (Ware, 1986: 131, quoted in Scott, 1990: 34).

Other organizations which offer 'opportunities' for participation in policy-making may in fact differ very little, apart from in a symbolic sense, from those organizations outlined above. Many public interest groups and other organizations within the 'new politics' genre offer what has been termed *astroturf* not *grassroots* opportunities for participation (Cigler and Loomis, 1995: 396).[14] This is clearly demonstrated by the comment by one FoE organizer in 1995 who said that: 'Members have to decide to back us or not. We make policy and if they don't like it they can join some other group.'

At best, decision-making in most large-scale groups can be termed *anticipatory rather than participatory democracy*, and more critically as *anticipatory oligarchy*. Decisions are made by the few on behalf of the mass within a framework which the *few* believe will be popular enough to maintain support. Such a 'pull' seems also to be at work in the public interest group field. Groups, such as FoE, find it to be an organizational necessity to lead from the front and hope that the members follow.

However, these protest businesses do not *naively* lead and *hope* that members will follow. As sensible businesses they have done their market research (e.g. via membership surveys) and know what members can 'live with'. As part of a regular exercise, the RSPB contacted its members in 1992 to discover if a strategy of widening the group's appeal as a broad environmental group could be pursued without eroding the traditional 'bird' constituency: group leaders anticipated potential exit. Rothenberg (1992: 10) observed that Common Cause gave top priority to electoral finance at the behest of its membership, via polls of members interests.[15] In fact, the discussion on the conflict between 'success' and 'influence' has even wider implications. For example, Judis (1992: 92 quoted in Bosso, 1995: 115) argues that groups may focus on what appeals to their source of income rather than what is important: 'None of the major organizations that rely on direct mail emphasize the redistribution of income, the rebuilding of cities, the rights of workers to join unions, the need for national health insurance, or the kinds of environmental issues that plague working-class neighborhoods.'

Thus public interest/campaigning/protest group politics do not significantly extend participatory democracy. But this is not a criticism of the groups because they have not set themselves up in the business of enhancing democracy. Groups such as Greenpeace and FoE are committed to maintaining a high profile for environmental ends: the mass membership is a tool in that process.

In arguing his case that there has been a decline in associational participation in the United States and thereby 'social capital,' Putnam has conceded that the decay of traditional forms of social organization may have been replaced by new organizations such as the Sierra Club or the National Organization for Women, or the American Association of Retired Persons which grew from 400,000 members in 1960 to 33 million in 1993. Putnam (1995: 71) concedes that these new mass membership organizations are of great political importance but he says that from the point of view of 'social connectedness' they are sufficiently different from classic 'secondary associations':

that we need to invent a new label – perhaps 'tertiary associations'. For the vast majority of their members, the only act of membership consists in writing a check for dues or perhaps occasionally reading a newsletter. Few ever attend any meetings of such organizations, and most are unlikely ever (knowingly) to encounter any other member.

Dalton (1994: 8, 19) specifically cites the FoE in Britain as an example of a new social movement(/organization). These organizations are connected with 'grassroots participation' and are held to enhance participatory democracy. But in our understanding of the 'members' few do more than make a small subscription and this adds to up to little in terms of significant participation. As Salisbury (1975: 186) argued: 'A large proportion of the expressive groups ... like retail merchants, offer customers the opportunity to patronize their enterprize but no chance to select either the clerks or the management or to do more than request a new line of goods.' There is little participatory democracy in protest business type organizations.

If the groups studied (FoE and Amnesty) have been part of a distinct new social movement network we would have expected them to be (at least) close to the stereotypical pressure group elaborated on in chapter 2 with opportunities for member participation, and a membership in control of a democratically accountable leadership. For example as Baer and Bositis (1993: 197) have argued, social movements can be seen as *an antidote to elite oligarchy*. However, the vast majority of the evidence about mail order groups is to the contrary. In our perspective the NSM sort of image of FoE conflates two sorts of environmental support. There may be an activist/lifestyle environmentalism for which the NSM imagery is appropriate, but the large-scale FoE is nearer something that can be termed a *protest business* than an NSM organization. 'Members' have minimalist (and primarily financial) obligations. There is skilled professional stimulation of support that borrows from business practice. There is no internal democracy. FoE members who disagree with (say) nuclear policy or wind-farms have no means to alter organizational policy. In Hirschman's language (1970), members' say in organizational policy is via *exit* not *voice*. Indeed, Amnesty and FoE (and we suspect other similar organizations, e.g. Greenpeace) seem just as affected by Michels' 'iron law of oligarchy' as are political parties.[16] As Berry (1977: 187) noted, the most interesting aspect about the organization of public interest groups is not that they are oligarchic in practice, but that there are not even formal concessions to a democratic structure in a majority of membership groups.

Shaiko (1993: 94) comments on Greenpeace (USA) as follows:

Whether or not this direct-mail strategy constitutes the dissemination of a new environmental ideology is open for debate. One might argue that it is antithetical to the tenets of the new environmental paradigm or postmaterialistic values as it is a method of economic manipulation perfected on Madison Avenue and permanently borrowed by the public interest sector in America.

Critiques of the 'pureness' of mass groups or protest businesses such as FoE or Greenpeace are to a large extent social science constructs. As argued above, the groups do not see themselves as being vehicles for the expansion of participatory democracy, and nor do the members themselves.[17] Table 7.1 shows that many members do not see members of FoE and Amnesty as a means of being 'active in political issues': 70.4 per cent of FoE members and 72.1 per cent of Amnesty members said it was not a 'very important reason'/'played no role whatsoever' in their decision to join. As Richardson (1995: 135) puts it, in the market for political activism individuals are prepared to contract out the participation task to organizations.

Table 7.1 *Being politically active as a reason for joining FoE/Amnesty International.*

Reason for joining	New (%)	Veteran (%)	Lapsed (%)	Weighted FoE sample (%)	Amnesty (%)
Very important	6.4	6.5	3.8	6.5	5.8
Important	21.7	19.8	19.3	20.1	18.5
Not very important	32.9	29.6	31.1	30.1	32.6
Played no role whatsoever	36.7	41.0	43.9	40.3	39.5

Rüdig *et al.* (1991: 40) found that in relation to the Green Party direct participation in the activities of the organization is a minority activity. The main contact between the members and the party is via the mail. Rüdig *et al.* (1991: 40) stated that their data tended 'to suggest that the Greens have a large body of passive members whose activity appears to be limited to the payment of the membership subscription'. Whiteley *et al.*'s (1994: 68–9) data on the Conservative Party membership, with regard to more active participation, suggests that it differs little from the Green Party, FoE, and Amnesty. For example, 68 per cent of 'members' said that they had not attended a party meeting in the previous year, and 83 per cent considered themselves to be not very active/not active. Whiteley *et al.* (1994: 75) argued that overall, the results indicate that for most party members, membership means donating money to the party on a regular basis, and little else. However, Seyd and Whiteley (1992: 88–9) found that 50 per cent of Labour Party members had devoted at least five hours of their time to 'party activities in the average month' and that 64 per cent had 'attended a Labour party meeting' in the last year. They consequently argued (1992: 88) that there is a lot more work done by party members than many observers of Labour Party politics suspected. Nevertheless, Seyd and Whiteley (1992: 202) concluded that: 'There is a feeling that the Labour party at the time of writing appears to be increasingly "de-energized", an organization in which members are increasingly passive rather than active, disengaged rather than engaged.'

Given the recent work that has been carried out on political party membership, it could be argued – somewhat ironically – that the thesis that participation through membership of public interest groups such as FoE and Amnesty is a direct replacement for participation through party membership has been strengthened – but only to the extent that it is equally slight in both cases. *People now appear to prefer to do very little in public interest groups as opposed to doing very little in political parties.* Ironically, political parties like the Labour Party in the United Kingdom in its modern guise are now even more open to the 'iron law of oligarchy'; it is also ironic that the so-called 'new politics' groups, whose philosophical underpinnings include a commitment to increasing participation, are even 'better-fits' to the generalization.

Hayes (1986: 142) argues these types of group 'clearly fall short of the pluralistic ideal in their internal decision-making'. He also claims that there is a conflict of interest between the leadership and the rank-and-file resulting from the life-threatening issue of organizational survival. Hayes (1986: 142) argues that leaders and followers will have divergent 'policy views'/interests, members seek the achievement of certain goals both tangible and utopian. Leaders essentially seek the utopian because: 'For group leaders, the ideal situation is rather to be perceived as making steady progress towards distant, intangible goals that are never quite achieved.'

The above discussion suggests that the relationship between the supporters of large 'mail order' groups and their organizations are some distance from the stereotype of pressure group membership where there is a policy-making role. The mail order groups are a means to allow the individual to relate to the political system and make a personal statement about concerns. Those who respond to appeals for finance are not concerned whether or not they are donors or obtaining some constitutional relationship.

In the large public interest groups we are considering there is little by way of political control: many organizations deliberately avoid democratic structures. The market research engaged upon by groups means that the group direction is steered to some extent by supporter/member attitudes. The group policy tends to be that which will maximize support. The 'role of such anticipated reactions' can be therefore be powerful in maintaining a link, however tenuous, between the leadership elite and membership 'supporters'. Writing about the different cases of the Moral Majority and Greenpeace, Godwin (1988: 48) has noted that all members need to do is: '*send in the money, the leaders will take the risks*' (emphasis added). Instead of conceiving these organizations as mass political bodies 'they' may be better represented as 'supplier/customer' relationships.

Major British organizations such as the RSPB, FoE, and Amnesty, are best seen as organizations with financial supporters rather than membership bodies. Participation through contribution is widespread. Many organizations attract financial supporters rather than members with a policy role; this practice circumvents the problems of internal democracy and policy interference. (For example, Greenpeace UK has a voting membership of between 20 and 30, which its Executive Director, Lord Melchett argues 'makes for fast decision-making' (*Daily Telegraph*, 22 June 1995).) We support the view that 'for many groups, the objective is a quiescent contributor, not an active member' (Godwin, 1988: 48).

Dunleavy (1991: 19) argues for the need for more research into why maintaining internal group democracy is so pervasive a strategy for leaders and groups. We prefer the term *internal responsive-*

ness rather than internal democracy. The goal of the leadership is often to remain responsive to its large number of quiescent subscribers: this falls some way short of internal democracy.

Many forms of democracy have been identified – with the main variants being direct participatory democracy and representative democracy. Insofar as public interest group membership is an alternative to representative democracy, then it appears that mass public interest group membership is a return to some of the virtues associated with direct participation. Our data does not support that interpretation and we see the political significance of mass group membership as marginal because the costs of such participation are so low that it cannot be assumed that members are making the commitment and investment implied by the direct participation model. Hayes (1986: 135) has presented these 'mass organizations' (in his terms) as potentially unsatisfactory democratic devices. He says that they become large, centralized, national groups lacking local subunits through which individual members can influence group decision-making. He goes on: 'Individuals within such organizations are faced with a remote and impersonal national leadership. Increasingly isolated and alienated, they become vulnerable to manipulation by group leaders.'

Godwin (1988: 144), however, argued that the *fears* that direct marketing would decrease democracy in America:

for the most part ... have not been realized. Direct marketing does not fragment the political parties; it does not replace democratic participation with ersatz participation where the public is manipulated by national elites; it does not increase significantly political extremism among the public ... [Neither has] Direct mail ... led to greater participation by minorities, the poor, and previously disadvantaged sectors of our society.

In fact it is the casualness of the connection between the revolving door membership and the organization that makes Hayes' assessment too pessimistic. There cannot be confidence that participation to the limited extent that is found in mail order groups is a significant democratic form of activity. But this expectation is one that has been a fence to jump set up by observers of these groups and not a selling point by the groups. They see the mass support as the means to facilitate protest rather than as policy-makers. Expectations ultimately derived from a social movement interpretation of the events, have led us to judge groups by their degree of success in changing the political system rather than simply (if importantly) changing political priorities.

This volume has argued that large-scale 'new politics' groups in the United Kingdom have added very little to the enhancement of participatory democracy associated with their characterization within a new social movement paradigm. Such a perception would be strongly associated with 'strong democracy' which 'rests on the idea of a self-governing community of citizens' and emphasizes *meaningful* participation (Barber (1984: 117). However, these groups provide a linkage to the political system which is much more characteristic of 'thin democracy', which according to Barber (1984: 24–5):

yields neither the pleasures of participation nor the fellowship of civic association, neither the autonomy of self governance of continuous political activity nor the enlarging mutuality of shared public goods of mutual deliberation, decision, and work ... thin democratic politics is at best a politics of static interests, never a politics of transformation; a politics of bargaining and exchange, never a politics of invention and creation ...

In some ways this book has been covering ground addressed by Dalton (1994) in *The Green Rainbow*. His interpretation is to see radical movements as acting in certain constrained ways because of their ideology. For example, he says that:

Ecologists' commitment to an alternative social paradigm should lead them to adopt a range of organizational and political behaviours that distinguish them from conservation groups ... The options open to a challenging group, for example, differ from those of a socially accepted organization such as the Civic Trust or the RSPB. Because of their identity and closer ties with the political establishment, conservation groups are more likely to engage in conventional political activities such as sitting on government commissions.

Our study found the differences to be far less dramatic. The influence of ideology is weakened as groups battle for influence. Dalton's own evidence causes him to concede important reservations to the thrust of his 'head line' argument. For example (1994: 252), he accepted that ecology groups participate in governmental consultation to a greater extent than the prior literature suggested. He also accepted that environmental interest groups were more likely to have a centralized structure than was expected. The price of effectiveness seems to be the acceptance of many aspects of traditional organization. Ingram, Colnic, and Mann (1995: 117) have noted that there is longing in some quarters for the emergence of a 'holistic' understanding of the environment. They say that encouraged by the success of 'green parties' and public opinion data some analysts of this persuasion see 'the transformation of fragmented, narrow, particularistic lobbies into a broad-scale social movement'. However, they found 'quite the opposite': 'the relationships among environmental interest groups have failed to coalesce into a unified environmental movement'. This is a conclusion we share.

This decision about how to characterize group participation is important. At the core of new work on participation is a sense that NSMs are the current fashion. Fuchs and Klingemann (1995: 18) suggest that NSMs can be distinguished from traditional collective actors in addressing new issues and providing different participation opportunities. They say:

> Thus non-institutionalized forms of participation are more appropriate for the expression of self-actualization values in the political arena ... Thus, with the emergence of new issue demands, one expects participation in elections to decline and involvement in non-institutionalized forms of political involvement to increase.

If, as we suggest, mail order groups are removed from the interpretative scope of new social movements, is the importance of the NSM phenomenon drastically reduced?

Notes

1   Berry (1984: 460) says the trends are independent but the growth of groups has benefited from party decline. In his view the alienation of the public from government both weakened the parties and gave groups an opportunity to expand their role.

2   Widfeldt (1995: 139) estimated that the percentage of the total electorate in membership of political parties in the United Kingdom fell from 9 per cent in 1964 to 3 per cent in 1987.

3   By late 1995 the Labour Party was claiming that its new-style membership drive had pushed the total up to near 400,000. The recent success of the Labour Party in increasing its membership base perhaps confirms the importance of recruitment technique. The new campaigns are borrowing the practices of the major groups.

4   Based on such a dramatic decline. Whiteley *et al.* (1994: 222) found that the membership of the Conservative Party had fallen from over 2 mil-

lion in the 1950s to 756,000 in 1994. Whiteley *et al.* (1994: 222) devised a model (which they conceded should be interpreted with care) on Conservative Party membership which estimated that the party 'has been losing 64,000 members each year since 1960. On the assumption that the model is a good predictor of future trends, it implies that if nothing is done to change this, the party membership will fall below 100,000 by the end of the century.'

5   In the United States Walker (1983) found that 30 per cent of citizen groups had formed since 1975 (to around.1982) and that by 1980 they comprised over 20 per cent of all groups represented in Washington.

6   The role of the large numbers of those supporting large-scale groups is financial contribution rather than personal activism. Johnson (1995:12) reports on a sample of sixty-one US environmental organizations and he found that only one-fifth offered members a chance to participate in 'political strategizing' or in meetings about the organizational future of the group.'

7   Similarly in the United Kingdom, Dearlove (1973: 169) argued that an aggressive 'improper' approach is likely to be 'prompted by the nature of their support base and the particular ideology of the group leaders'.

8   The Channel Four documentary *Where Have the Warriors Gone?* noted that Greenpeace's work with refrigerator manufacturers in Germany to get a switch away from the use of ozone-damaging CFCs as refrigerants marked a significant transformation in the organizational *modus operandi*. Previously Greenpeace 'never "got into bed" with industry, politics or any other vested interest, now [it is] dabbling in the white-goods trade like any shopkeeper or commercial traveller' (Greenpeace booklet, 1994: 11).

9   This development may have gained a greater momentum because of the sectorization of policy-making. As Richardson (1995: 127) argues: 'By concentrating on specific issues, the individual can join a group of genuinely like-minded people, can generally avoid the typical left/right ideological splits common in parties and, moreover, may see more immediate policy results.'

10   See Dickson and McCulloch (1996) for a fuller discussion of the Brent Spar campaign.

11   For example, Greenpeace USA has been attacked by critics of direct mail and other environmentalists because its *action agenda* 'is shaped powerfully by whatever new eco-crisis (for example, the killing of dolphins in tuna fishing) it can exploit in the millions of pieces of mail to send out' (Bosso, 1995: 111).

12   As Webb (1994: 118) points out, it is worth noting that the Conservative Party has never made any pretence of running a democratic organization. He says that the party developed out of a cadre-type organization based initially around parliamentary elites, noting that formally, all author-

ity and policy emanates from the leader, and the organization exists primarily to recruit and to aid the leadership. The membership is formally separate from both the parliamentary party and the extra-parliamentary party's organizational headquarters. The annual conference itself is merely a (highly stage-managed) consultation process, as well as a morale-boosting exercise for the activists.

13   Webb (1994: 112) also says that: 'It is perhaps not too great an exaggeration to say that, at least for Labour and the Conservatives, individual members are simply not crucial to the parties, in either an electoral or in a financial sense.' Individual membership in the 1990s is a smaller part of total membership than in the 1960s: 'in 1960, it formed 12.5 per cent of total, falling to 10.9 per cent in 1970 and to just 5.7 per cent in 1992' (Webb, 1994: 114).

14   In 1995 Greenpeace was reported to be trying to change its supporters into campaigners rather than mere fund-raisers. This was presented in the press as a decision that was taken in spite of its financial implications: 'the board wants them to take part in mass protest actions, such as the recent (anti nuclear) demonstration outside Chequers when John Major met the French President, Jacques Chirac' (*Independent*, 6 November 1995). However, this was hardly a signal that Greenpeace was to be a wholeheartedly participatory body: at the same time the budget for advertising and direct mail was to be increased by £100,000. In fact, initially, the decision to campaign more can be seen to be driven by its fund-raising needs. This is the 'unique selling proposition' of the organization and this sort of radical image is itself a factor in the joining decisions of the more passive supporters

15   Rothenberg (1992: 27) records that the leadership selects its policy areas carefully to maintain membership loyalty: 'The staff employs polls and other less formal mechanisms to gather data on constituents' preferences .'

16   Both McKenzie (1964) and Minkin (1980) found that with regard to internal party democracy that the Labour and Conservative parties differed little. The leadership of both parties dominated the policy-making process.

17   We follow Parry *et al.* (1992: 66) in defining participation not in terms of 'mere membership' of interest groups, political parties or NSMs, but as 'based on actions, on doing things which are intended to have some effect on political outcomes ... It is perfectly possible to join a group or a party or trade union and to play an entirely passive role.'

References

Aarts, K. (1995), 'Intermediate Organizations and Interest Representation', in H.-D. Klingemann and D. Fuchs (eds), *Citizens and the State* (Oxford: Oxford University Press).

Almond, G. and Verba, S. (1963), *The Civic Culture. Political Attitudes and Democracy* (Boston: Little, Brown and Company).

Amenta, E. and Skocpol, T. (1989), 'Taking Exception: Explaining the Distinctiveness of American Public Policies in the Last Century', in F. Castles, *The Comparative History of Public Policy* (New York: Oxford University Press).

Amnesty International (1993), *The 1993 Report on Human Rights Around the World* (Alameda, CA: Amnesty International).

Axlerod, R. (1984), *The Evolution of Cooperation* (New York: Basic Books).

Baer, D. L. and Bositis, D. A. (1993), *Politics and Linkage in a Democratic Society* (Englewood Cliffs, NJ: Prentice Hall).

Barber, B. (1984), *Strong Democracy* (Berkeley: University of California Press).

Barho, R. (1986), *Building the Green Movement* (London: Heretic Books).

Barry, B. (1970), *Sociologists, Economists and Democracy* (London: Collier-Macmillan).

Baumgartner, F. and Walker, J. (1988), 'Survey Research and Membership in Voluntary Associations', *American Journal of Political Research*, 32: 908–28.

Bennie, L. G. and Rüdig, W. (1993), 'Youth and the Environment: Attitudes and Action in the 1990s', *Youth and Policy*, 42: 6–21.

Berry, J. M. (1977), *Lobbying for the People* (Princeton, NJ: Princeton University Press).

— (1984), *The Interest Group Society* (Glenview, IL: Scott, Foresman/Little Brown).

— (1993), 'Citizen Groups and the Changing Nature of Interest Group Politics in America', *The Annals*, AAPSS, 528: 30–41.

Bomberg, E. (1993), 'The German Greens and the European Community: Dilemmas of a Movement-Party', in D. Judge (ed.), *Environmental Politics (Special Issue) – A Green Dimension for the European Community: Political Issues and Processes*, 1(4): 160–85.

Bosso, C. J. (1991), 'Adaptation and Change in the Environment Movement', in A. J. Cigler and B. A. Loomis (eds), *Interest Group Politics*, 3rd edn (Washington, D.C.: Congressional Quarterly Press).

— (1994), 'After the Movement: Environmental Activism in the 1990s' in N. Vig and M. Kraft (eds), *Environmental Policy in the 1990s*, 2nd edn (Washington, D.C.: Congressional Quarterly Press).

— (1995), 'The Color of Money: Environmental Groups and the Pathologies of Fund Raising', in A. J. Cigler and B. A. Loomis (eds), *Interest Group Politics*, 4th edn (Washington, D.C.: Congressional Quarterly Press).

Bridgwood, A. and Savage, D. (1991), *1991 General Household Survey* (London: HMSO).

Browne, W. P. (1991), 'Issue Niches and the Limits of Interest Group Influence', in A. J. Cigler and B. A. Loomis (eds), *Interest Group Politics*, 3rd edn (Washington, D.C.: Congressional Quarterly Press).

Burke, T. (1982), 'Friends of the Earth and the Conservation of Resources', in P. Willetts (ed.), *Pressure Groups in the Global System: The Transnational Relations of Issue-Orientated Non-Governmental Organizations* (London: Frances Pinter).

Burstein, P. (1995), 'What Do Interest Groups, Social Movements, and Political Parties Do? A Synthesis', a paper presented at the Annual Meeting of the American Political Science Association, Chicago, 2 September, 1995.

Butler, D. and Kavanagh, D. (1992), *The British General Election of 1992* (London: Macmillan).

Byrne, P. (1988), *The Campaign for Nuclear Disarmament* (London: Croom Helm).

Carson, R. (1965), *Silent Spring* (Harmondsworth: Penguin).

Central Statistical Office (CSO) (1992), *Annual Abstract of Statistics 1992* (London: HMSO).

Cigler, A. J. and Hansen J. M. (1983), 'Group Formation Through Protest' in A. J. Cigler and B. A. Loomis (eds), *Interest Group Politics* (Washington D.C., Congressional Quarterly Press).

Cigler, A. J. and Loomis, B. A. (1995), 'Contemporary Interest Group Politics: More Than "More of the Same"', in A. J. Cigler and B. A. Loomis (eds), *Interest Group Politics*, 4th edn (Washington, D.C.: Congressional Quarterly Press).

Clark P. B. and Wilson, J. Q. (1961), 'Incentive Systems: A Theory of Organizations', *Administrative Science Quarterly*, 6: 129–66.

Cohen, J. L. (1985), 'Strategy and Identity: New Theoretical Paradigms and Contemporary Social Movements', *Social Research*, 52: 663–716.

Cohen, S. (1995), *Denial and Acknowledgement: The Impact of Information About Human Rights Violations* (Jerusalem: University of Jerusalem).

Commoner, B. (1963), *Science and Survival* (New York: Viking Books).

Cook, C. E. (1984), 'Participation in Public Interest Groups', *American Politics Quarterly*, 12: 409–30.

Corsino, L. (1986), 'Campaigning Organizations, Social Technology and Apolitical Participation', *New Political Science*, 14: 141–55.

Costain, W. D. and Lester, J. P. (1995), 'The Evolution of Environmentalism', in J. P. Lester (ed.), *Environmental Politics and Policy*, 2nd edn (Durham and London: Duke University Press).

Cotgrove, S. (1982), *Catastrophe of Cornucopia* (New York: John Wiley and Sons).

Cowell, S. (1990), *Who's Who in the Environment: England* (London: The Environment Council).

Crenson, M. A. (1987), 'The Private Stake in Public Goods: Overcoming the Illogic of Collective Action', *Policy Sciences*, 20: 259–76.

Dalton, R. J. (1993), 'Preface', *The Annals*, AAPSS, 528: 8–12.

— (1994), *The Green Rainbow* (New Haven: Yale University Press).

Dalton, R. J., Kuechler, M. and Bürklin, W. (1990), 'The Challenge of New Movements', in R. J. Dalton and M. Kuechler (eds), *Challenging the Political Order: New Social and Political Movements in Western Democracies* (Oxford: Polity Press).

Dearlove, J. (1973), *The Politics of Policy in Local Government* (Cambridge: Cambridge University Press).

Dickson, L. and McCulloch, A. (1996), 'Shell, The Brent Spar and Greenpeace: A Doomed Tryst?', *Environmental Politics*, 5(1): 122–9.

Dowding, K. (1991), *Rational Choice and Political Power* (Aldershot: Edward Elgar).

Downs, A. (1972), 'Up and Down with the Ecology', *Public Interest*, 28: 38–50.

Dunlap, R. (1992), 'Public Opinion and Environmental Policy', in J. P. Lester (ed.), *Environmental Politics and Policy*, 2nd edn (Durham and London: Duke University Press).

Dunleavy, P. (1991), *Democracy, Bureaucracy and Public Choice* (London: Harvester Wheatsheaf).

Elster, J. (1979), *Ulysses and Sirens. Studies in Rationality and Irrationality* (Cambridge: Cambridge University Press).

— (1989), *The Cement Society* (Cambridge: Cambridge University Press).

Ennals, M. (1982), 'Amnesty International and Human Rights', in P. Willetts (ed.), *Pressure Groups in the Global System: The Transnational Rela-*

tions of Issue-Orientated Non-Governmental Organizations (London: Frances Pinter).

Eulau, H. (1989), 'Crossroads of Social Science', in H. Eulau (ed.), *Crossroads of Social Science: The ICPSR 25th Anniversary Papers* (New York: Agathon Press).

Evan, G., Taylor, B. and Heath, A. (1994), 'Class, Vote and (Postmaterialist) Values: Examining the Impact of the New Agenda on Class Politics in Britain', *Centre for Research into Elections and Social Trends (CREST) Working Paper Series, No. 29* (London: CREST).

Eyerman, R. and Jamison, A. (1989), 'Social Movements: Contemporary Debates', *Research Report, Department of Sociology, Lund University* (Lund: Lund University).

Featherstone, M. (1991), *Consumer Culture and Postmodernism* (London, Sage).

Ferejohn, J. (1991), 'Rationality and Interpretation: Parliamentary Elections in Early Stuart England', in K. R. Munro (ed.), *The Economic Approach to Politics: A Critical Reassessment of the Theory of Rational Action* (New York: Harper Collins).

Ferree, M. (1992), 'The Political Context of Rationality,' in A. Morris and C. M. Mueller (eds), *Frontiers in Social Movement Theory* (New Haven: Yale University Press.).

Finkel, S., Muller, E. and Opp, K.-D. (1989), 'Personal Influence, Collective Rationality and Mass Political Action', *American Political Science Review*, 83: 885–904.

Flynn, A. and Lowe, P. (1992), 'The Greening of the Tories: The Conservative Party and the Environment', in W. Rüdig (ed.), *Green Politics Two* (Edinburgh: Edinburgh University Press).

Franklin, M. N. and Rüdig, W. (1993), 'The Green Voter in the 1989 European Elections' in D. Judge (ed.), *Environmental Politics (Special Issue) – A Green Dimension for the European Community: Political Issues and Processes*, 1(4): 129–59.

Frisch, M. (1994), *Directory for the Environment* (London: Merlin Press).

Fuchs, D. and Klingemann, H.-D. (1995), 'Citizens and the State: A Changing Relationship' in H.-D. Klingemann and D. Fuchs (eds), *Citizens and the State* (Oxford: Oxford University Press).

Gallagher, M., Laver, M. and Mair, P. (1992), *Representative Government in Western Europe* (New York: McGraw-Hill).

Gamson, W. (1975), *The Strategy of Social Protest* (Homewood: Dorsey Press).

Gelb, J. (1989), *Feminism and Politics: A Comparative Perspective* (Berkeley: University of California Press).

Godwin, R. K. (1988), *One Billion Dollars of Influence* (Chatham House, NJ: Chatham House).

— (1992), 'Money, Technology, and Political Interests: The Direct Marketing of Politics', in M. Petracca (ed.), *The Politics of Interests* (Colorado: Westview Press).

Godwin, R. K. and Mitchell, R. C. (1982), 'Rational Models, Collective Goods and Nonelectoral Political Behaviour', *Western Political Quarterly*, 35: 161–92.

Goldsteen, R. and Schorr, J. (1991), *Demanding Democracy after Three Mile Island* (Gainesville: University of Florida Press)

Grafstein, R. (1994), 'The Concept of Rationality', paper presented at the 16th World Congress of the International Political Science Association, August, Berlin.

Grant, Wyn (1978), 'Insider Groups, Outsider Groups and Interest Group Strategies in Britain', University of Warwick Department of Politics, *Working Paper no. 19*.

— (1995), *Pressure Groups, Politics and Democracy in Britain*, 2nd edn (Hemel Hempstead: Harvester Wheatsheaf).

Gray, V. and Lowery, D. (1993), 'The Diversity of the Interest Group Systems', *Political Research Quarterly*, 46(1): 81–97.

Green, D. P. and Shapiro, I. (1994), *The Pathologies of Rational Choice Theory* (New Haven: Yale University Press).

Grove-White, R. (1992), 'GrossBritannien', in C. Hey, U. Brendle and C. Weinber (eds), *Umweltverbande und EG* (Frieburg: EURES – Institut für regionale Studien).

Gundelach, P. (1995), 'Grass-Roots Activity', in J. W. van Deth and E. Scarbrough (eds), *The Impact of Values* (Oxford: Oxford University Press): 412–40.

Habermas, J. (1976), *Legitimation Crisis* (London: Heinemann).

Hager, C. (1993), 'Citizen Movements and Technological Policy Making in Germany', *The Annals, AAPSS*, 528: 42–55.

Hansen, J. M. (1985), 'The Political Economy of Group Membership', *American Political Science Review*, 79: 79–96.

Hardin, R. (1979), 'Comments', in C. Russell (ed.), *Collective Decision-Making* (Baltimore: Johns Hopkins University Press): 122–9.

Hayes, M. T. (1986), 'The New Group Universe', in A. J. Cigler and B. A. Loomis (eds), *Interest Group Politics*, 2nd edn (Washington, D.C.: Congressional Quarterly Press): 133–45.

Hershey, M. R. (1993), 'Citizens' Groups and Political Parties in the United States', *The Annals, AAPSS*, 528: 142–56.

Hirschman, A. (1970), *Exit, Voice and Loyalty: Responses to Decline in Firms, Organizations and States* (Cambridge, MA: Harvard University Press).

— (1982), *Shifting Involvements* (Oxford: Martin Robertson).

Horton, T. (1991), 'The Green Giant', *Rolling Stone*, 5 September: 44.

Imig, D. (1994), 'Advocacy by Proxy: The Children's Lobby in American Politics', a paper prepared for the panel on 'Interest Representation, Issue Networks, and Policymaking in the US', at the Annual Meeting of the American Political Science Association, New York, 1–4 September.

Inglehart, R. (1977), *The Silent Revolution: Changing Values and Political Styles among Western Publics* (Princeton, NJ: Princeton University Press).

— (1981), 'Post-Materialism in an Environment of Insecurity', *American Political Science Review*, 85: 880–900.

— (1990), 'Values, Ideology, and Cognitive Mobilization in New Social Movements', in R. J. Dalton and M. Kuechler (eds), *Challenging the Political Order: New Social and Political Movements in Western Democracies* (Cambridge: Polity Press).

— (1995), 'Public Support for Environmental Protection: Objective Problems and Subjective Values in 43 Societies', *PS: Political Science and Politics*, March: 57–72.

Inglehart, R. and Klingemann, H. D. (1979), 'Ideological Conceptualization and Value Priorities', in S. H. Barnes, M. Kaase, *et al.* (eds), *Political Action: Mass Participation in Five Western Democracies* (Beverley Hills, CA: Sage).

Ingram, H., Colnic, D. and Mann, D. (1995), 'Interest Groups and Environmental Policy,' in J. P. Lester (ed.), *Environmental Politics and Policy*, 2nd edn (Durham and London: Duke University Press).

Jacobs, E. and Worcester, R. (1990), *We British: Britain Under the Moriscope* (London: Weidenfeld and Nicolson).

Jahn, D. (1993), 'The Rise and Declines of New Politics and the Greens in Sweden and Germany: Resource Dependence and New Social Cleavages', *European Journal of Political Research*, 24: 177–94.

Jamison, A., Eyerman, R. and Cramer, J. (1990), *The Making of the New Environmental Consciousness: A Comparative Study of the Environmental Movements in Sweden, Denmark and the Netherlands* (Edinburgh: Edinburgh University Press).

Jeffrey, R. C. (1983), *The Logic of Decision* (Chicago: University of Chicago Press).

Jelen, T. G., Thomas, S. and Wilcox, C. (1994), 'The Gender Gap in Comparative Perspective: Gender Differences in Abstract Ideology and Concrete Issues in Western Europe', *European Journal of Political Research*, 25(2): 171–86.

Johnson, P. E. (1995), 'How Environmental Groups Recruit Members: Does the Logic Still Hold Up?', a paper presented at the Annual Meeting of the American Political Science Association, Chicago, 2 September1995.

Jordan, G., Maloney, W. A. and McLaughlin, A. M. (1992), 'What is Studied when Pressure Groups are Studied?', *British Interest Group Project*,

Working Paper Series no. 1 (Aberdeen: British Interest Group Project, University of Aberdeen).

Jowell, R., Witherspoon, S. and Brook, L. (1990), *British Social Attitudes: The 7th Report* (Aldershot: Gower).

Judis, J. B. (1992), 'The Pressure Elite: Inside the Narrow World of Advocacy Group Politics', *The American Prospect* (Spring).

Kasse, M. (1990), 'Social Movements and Political Innovation', in R. J. Dalton and M. Kuechler (eds), *Challenging the Political Order: New Social and Political Movements in Western Democracies* (Oxford: Polity Press).

Kimber, R. (1981), 'Collective Action and the Fallacy of the Liberal Fallacy', *World Politics*, 33(2): 178–96.

Kimber, R. and Richardson, J. J. (1974), *Pressure Groups in Britain* (London: Dent).

King, D. and Walker, J. L. (1992), 'The Provision of Benefits by Interest Groups in the United States', *The Journal of Politics*, 54: 394–426.

Kitschelt, H. (1993), 'Social Movements, Political Parties and Democratic Theory', *The Annals, AAPSS*, 528: 13–29.

Knoke, D. and Wisely, N. (1990), 'Social Movements' in D. Knoke, *Political Networks* (Cambridge, Cambridge University Press)

Kriz, M. (1990), 'Shades of Green', *National Journal*, 28 July: 1826–31.

Kuechler, M. and Dalton, R. (1990), 'New Social Movements and the Political Order: Inducing Change for Long-term Stability', in R. J. Dalton and M. Kuechler (eds), *Challenging the Political Order: New Social and Political Movements in Western Democracies* (Oxford: Polity Press).

Lawson, K. and Merkl, P. (eds) (1988), *When Parties Fail: Emerging Alternative Organizations* (Princeton: Princeton University Press).

Lester, J. P. (ed.) (1995), *Environmental Politics and Policy*, 2nd edn (Durham and London: Duke University Press).

Letto, J. (1992), 'One Hundred Years of Compromise', *Buzzworm*, 4 (March/April).

Lidskog, R. (1994), *On Studying Social Movements*, Working Paper, University of Örebro.

Lipow, A. and Seyd, P. (1996), 'The Politics of Anti-Partyism', *Parliamentary Affairs*, 49(2): 273–84.

Loomis, B. A. and Cigler, A. J. (1986), 'Introduction: The Changing Nature of Interest Group Politics', in A. J. Cigler and B. A. Loomis (eds), *Interest Group Politics*, 2nd edn (Washington, D.C.: Congressional Quarterly Press).

— (1991), 'Introduction: The Changing Nature of Interest Group Politics', A. J. Cigler and B. A. Loomis (eds), *Interest Group Politics*, 3rd edn (Washington, D.C.: Congressional Quarterly Press).

Lowe, P. and Goyder, J. (1983), *Environmental Groups in Politics* (London:

Allen & Unwin).

Lowe, P. and Rügid, W. (1986), 'Political Ecology and the Social Sciences', *British Journal of Political Science*, 16: 513–50.

Lowery, D. and Gray, V. (1992), 'The Density of State Interest Group Systems', *Journal of Politics*, 54(4): 191–206.

— (1995), 'The Population Ecology of Gucci, Gulch, or the Natural Regulation of Interest Group Numbers in the American States', *American Journal of Political Science*, 39(1): 1–29.

McAdam, D. (1988), *Freedom Summer* (New York: Oxford University Press).

McAdam, D. and Rucht, D. (1993), 'The Cross National Diffusion of Movement Ideas', *The Annals, AAPSS* (July): 56–74.

McCarthy, J. and Zald, M. (1973), *The Trend of Social Movements in America* (Morristown: General Learning Press).

— (1977), 'Resource Mobilization and Social Movements' *American Journal of Sociology*, 82: 1112–41.

McCormick, J. (1991), *British Politics and the Environment* (London: Earthscan).

McFarland, A. S. (1976), *Public Interest Lobbies: Decision Making on Energy* (Washington D.C.: American Enterprise Institute).

— (1984), *Common Cause* (Chatham, NJ: Chatham House).

McKenzie, R. (1964), *British Political Parties*, 2nd edn (London: Mercury).

Maloney, W. A., Jordan, G. and McLaughlin, A. M. (1994), 'Interest Groups and Public Policy: The Insider/Outsider Model Revisited', *Journal of Public Policy*, 14(1): 17–38.

March, J. and Olsen, J. P. (1989), *Rediscovering Institutions* (New York, Free Press).

Margolis, H. (1982), *Selfishness, Altruism, and Rationality* (New York: Cambridge University Press).

Marsh, D. (1976), 'On Joining Interest Groups: An Empirical Consideration of the Works of Mancur Olson', *British Journal of Political Science*, 6: 257–71.

— (1994), 'Coming of Age, But Not Learning By Experience: One Cheer for Rational Choice Theory', paper presented at the Workshop on Rational Actor Models of Political Participation, Joint Session of the European Consortium for Political Research, Madrid, Spain, 17–22 April 1994.

Martell, L. (1994), *Ecology and Society: An Introduction* (Cambridge: Polity Press).

Marwell, G. and Ames, R. E. (1979), 'Experiments on the Provision of Public Goods: Resources, Group Size, and the Free-rider Problem', *American Journal of Sociology*, 48: 1335–60.

— (1981), 'Economists Free-ride, Does Anyone Else?', *Journal of Public Economics*, 15: 295–310.

Maslow, A. H. (1943), 'A Theory of Human Motivation', in J. M. Shafritz and A. C. Hyde (eds), *Classics of Public Administration*, 3rd edn (California: Brooks/Cole).

Melucci, A. (1989), *The Nomads of the Present* (London, Radius).

Michels, R. (1919), *Political Parties: A Sociological Study of the Oligarchical Tendencies of Modern Democracy* (New York: Free Press).

Minkin, L. (1980), *The Labour Party Conference: A Study in the Politics of Intra-party Democracy* (Manchester: Manchester University Press).

Mitchell, R. C. (1979), 'National Environmental Lobbies and the Apparent Illogic of Collective Action', in C. Russell (ed.), *Collective Decision-Making* (Baltimore: Johns Hopkins University Press).

Mitchell, R. C., Mertig, A. G. and Dunlap, R. E. (1991), 'Twenty Years of Environmental Mobilization: Trends Among National Environmental Organizations', in R. E. Dunlap and A. G. Mertig (eds), *American Environmentalism: The US Environmental Movement, 1970–1990* (Washington: Taylor and Francis).

Moe, T. M. (1980a), 'A Calculus of Group Membership', *American Journal of Political Science*, 24: 593–632.

— (1980b), *The Organization of Interests* (Chicago: Chicago University Press).

— (1981), 'Toward a Broader View of Interest Groups', *The Journal of Politics*, 43: 531–43.

Morris, A. D. and Mueller, C. M. (eds), (1992), *Frontiers in Social Movement Theory* (New Haven: Yale University Press).

Morrison, D. (1979), 'Uphill and Downhill Battles and Contributions to Collective Action', in C. Russell (ed.), *Collective Decision-Making* (Baltimore: Johns Hopkins University Press): 130–3.

Muller, E. and Opp, K.-D. (1986), 'Rational Choice and Rebellious Collective Action', *American Political Science Review*, 80: 471–87.

Mundo, P. A. (1992), *Interest Groups: Cases and Characteristics* (Chicago: Nelson Hall).

Nas, M. (1995), 'Green, Greener, Greenest', in J. W. van Deth and E. Scarbrough (eds), *The Impact of Values* (Oxford: Oxford University Press): 275–300.

Nowes, A. J. and Neeley, G. (1996), 'Public Interest Group Entrepreneurship and Theories of Group Mobilization', *Political Research Quarterly*, 49(1), pp. 119–46.

Offe, C. (1985), 'New Social Movements: Challenging the Boundaries of Institutional Politics', *Social Research*, 52(4): 818–68.

— (1990), 'Reflections on the Institutional Self-transformation of Movement Politics', in R. J. Dalton and M. Kuechler (eds), *Challenging the Political Order: New Social and Political Movements in Western Democracies* (Oxford: Polity Press).

Olson, M. (1965), *The Logic of Collective Action* (Cambridge, MA: Harvard University Press) (2nd edn 1971).

— (1979), 'Epilogue: Letter to Denton Morrison', *Research in Social Movements, Conflicts and Change*, 2: 149–50.

Opp, K.-D. (1986), 'Soft Incentives and Collective Action', *British Journal of Political Science*, 16: 87–112.

Parry, G., Moyser, G. and Day, N. (1992), *Political Participation and Democracy in Britain* (Cambridge: Cambridge University Press).

Peterson, M. (1992), 'Interest Mobilization and the Presidency', in M. P. Petracca (ed.), *The Politics of Interests* (Colorado: Westview).

Petracca, M. (1991), 'The Rational Choice Approach to Politics: The Challenge to Normative Democratic Theory', *Review of Politics*, 53: 289–319.

Poguntke, T. (1993), *Alternative Politics: The German Green Party* (Edinburgh: Edinburgh University Press).

Power, J. (1981), *Amnesty International: The Human Rights Story* (Oxford: Pergamon Press).

Putnam, R. D. (1995), 'Bowling Alone: America's Declining Social Capital', *Journal of Democracy*, 6(1): 65–78.

Rawcliffe, P. (1992), 'Swimming with the Tide – Environmental Groups in the 1990s', *ECOS*, 13(1): 2–9.

Richardson, J. J. (1995), 'The Market for Political Activism: Interest Groups as a Challenge to Political Parties', *West European Politics*, 18(1): 116–39.

Robinson, M. (1992), *The Greening of British Party Politics* (Manchester: Manchester University Press).

Rochon, T. R. and Mazmanian, D. A. (1993), 'Social Movements and the Policy Process', *The Annals, AAPSS*, 528: 75–87.

Roha, R. R. (1990), 'Giving Back', *Changing Times* (November).

Rohrschneider, R. (1988), 'Citizen Attitudes towards Environmental Issues: Selfish or Selfless?', *Comparative Politics*, 21: 347–67.

Rosenstone, S. and Hansen, J. M. (1993), *Mobilization, Participation, and Democracy in America* (New York: Macmillan).

Rothenberg, L. S. (1988), 'Organizational Maintenance and the Retention Decision in Groups', *American Political Science Review*, 82(4): 1129–52.

— (1992), *Linking Citizens to Government* (New York: Cambridge University Press).

Rucht, D. (1990), 'The Strategies and Action Repertoires of New Movements', in R. J. Dalton and M. Kuechler (eds), *Challenging the Political Order: New Social and Political Movements in Western Democracies* (Oxford: Polity Press).

Rüdig, W. (1995), 'Between Moderation and Marginalisation: Environmental Radicalism in Britain', in B. Taylor (ed.), *Ecological Resistance*

Movements: The Global Emergence of Radical and Popular Environmentalism (Albany, New York: State University of New York Press).

Rüdig, W., Bennie, L. G. and Franklin, M. N. (1991), *Green Party Members: A Profile* (Glasgow: Delta).

Rüdig, W., Franklin, M. N. and Bennie, L. G. (1993), *Green Blues: The Rise and Decline of the British Green Party*, Strathclyde Papers on Government and Politics no. 95 (Glasgow: Department of Government, University of Strathclyde).

Sabatier, P. A. (1992), 'Interest Group Membership and Organization: Multiple Theories', in M. Petracca (ed.), *The Politics of Interests* (Colorado: Westview Press).

Sabatier P. A. and McLaughlin, S. (1990), 'Belief Congruence Between Interest Group Leaders and Members', *Journal of Politics*, 52: 914–35.

Salazar, D. (1995), 'More Empty Cores: Interest Groups and Environmental Politics in Washington State', paper prepared for presentation at the American Political Science Association.

Salisbury, R. H. (1969), 'An Exchange Theory of Interest Groups', *Midwest Journal of Political Science*, 13: 1–32.

— (1975), 'Interest Groups', in F. I. Greenstein and N. W. Polsby (eds), *Nongovernmental Politics: Handbook of Political Science*, IV.

Samstag, T. (1988), *For Love of Birds: The Story of the Royal Society for the Protection of Birds* (Sandy, Bedfordshire: RSPB).

Scarbrough, E. (1995), 'Materialist-Postmaterialist Value Orientations', in J. W. van Deth and E. Scarbrough (eds), *The Impact of Values* (Oxford: Oxford University Press).

Schlesinger, A. (1986), *The Cycles of American History* (Boston: Houghton Mifflin).

Schlozman, K. L., Burns, N. Verba, S. and Donahue, J. (1995a), 'Gender and Citizen Participation: Is There a Different Voice?', *American Journal of Political Science*, 39(2): 267–93.

Schlozman, K. L., Verba, S. and Brady, H. E. (1995b), 'Participation's Not a Paradox: The View from American Activists', *British Journal of Political Science*, 25(1): 1–36.

Schlozman, K. and Tierney, D. (1986), *Organized Interests and American Democracy* (New York: Harper and Row).

Scott, A. (1990), *Ideology and the New Social Movements* (London: Unwin Hyman).

Scott, D. and Willits, F. K. (1994), 'Environmental Attitudes and Behavior: A Pennsylvania Survey', *Environment and Behavior*, 26: 88–100.

Sen, A. (1967), 'Isolation, Assurance and the Social Rate of Discount, *Quarterly Journal of Economics*, 81: 112–24.

— (1979), 'Rational Fools', in F. Hahn and M. Hollis (eds), *Philosophy and Economic Theory* (Oxford: Oxford University Press) .

Seyd, P. and Whiteley, P. F. (1992), *Labour's Grass Roots: The Politics of Party Membership* (Oxford: Clarendon Press).

Shaiko, R. G. (1991), 'More Bang for the Buck: The New Era of Full-Service Public Interest Organizations', in A. J. Cigler and B. A. Loomis (eds), *Interest Group Politics*, 3rd edn (Washington, D.C.: Congressional Quarterly Press).

— (1993), 'Greenpeace, USA: Something Old, New, Borrowed', *The Annals, AAPSS*, 528: 88–100.

Simon, H. (1993), 'The State of American Political Science: Professor Lowi's View of our Discipline', *PS, Political Science and Politics* (March): 49–51.

Smelser, N. J. (1962), *Theory of Collective Behaviour* (New York: Free Press).

Stanfield, R. (1985), 'Environmental Lobby's Changing of the Guard is Part of the Movement's Evolution', *National Journal*, 8 June: 1350–3.

Stern, P. C., Deitz, T. and Kalof, L. (1993), 'Value Orientations, Gender, and Environmental Concern', *Environment and Behavior*, 25: 322–48.

Taylor, M. (1976), *Anarchy and Cooperation* (London: John Wiley).

— (1988), *Rationality and Revolution* (New York; Cambridge University Press).

Thompson, G. P. (1985), 'The Environment Movement Goes Back to Business School', *Environment*, 27(4): 7–11, and 30.

Toch, H. (1965), *The Social Psychology of Social Movements* (London: Methuen).

Togeby, L. (1993), 'Grass Roots Participation in the Nordic Countries', *European Journal of Political Research*, 24: 159–75.

Tonge, J. (1994), 'The Anti-Poll Tax Movement: A Pressure Movement?', *Politics*, 14(2): 93–9.

Truman, D. B. (1951), *The Governmental Process: Public Interests and Public Opinion* (New York: Alfred A. Knopf).

Udehn, L. (1993), 'Twenty-five Years of *The Logic of Collective Action*', *Acta Sociologica*, 36: 239–61.

— (1996), *The Limits of Public Choice* (London: Routledge).

Van Liere, K. D. and Dunlap, R. E. (1980), 'The Social Bases of Environmental Concern', *Public Opinion Quarterly*, 44: 181–97.

Verba, S. and Nie, N. (1972), *Participation in America: Political Democracy and Social Equality* (New York: Harper and Row).

Walker, J. L. (1983), 'The Origins and Maintenance of Interest Groups in America', *American Political Science Review*, 77: 390–406.

— (1991), *Mobilizing Interest Groups in America* (Ann Arbor: University of Michigan Press).

Ware, A. (1996), *Political Parties and Party Systems* (Oxford: Oxford University Press).

Webb, P. (1994), 'Party Organizational Change in Britain: The Iron Law of Centralization?', in R. S. Katz and P. Mair (eds), *How Parties Organize* (London: Sage).

Whiteley, P. and Seyd, P. (1996a), 'Introduction: Rationality and political decision-making', *European Journal of Political Research*, 29(2): 143–5.

— (1996b), 'Rationality and Party Activism: Encompassing Test of Alternative Models of Political Participation', *European Journal of Political Research*, 29(2): 215–34.

Whiteley, P., Seyd, P. and Richardson, J. (1994), *True Blues: The Politics of Conservative Party Membership* (Oxford: Clarendon Press).

Widfeldt, A. (1995), 'Party Membership and Party Representativeness', in H.-D. Klingemann and D. Fuchs (eds), *Citizens and the State* (Oxford: Oxford University Press).

Wilson, G. (1990), *Interest Groups* (Oxford: Blackwell).

Wilson, J. Q. (1973), *Political Organizations* (New York: Basic Books).

Witherspoon, S. (1994), 'The Greening of Britain: Romance and Rationality', in R. Jowell, J. Curtice, L. Brook, D. Ahendt, and A. Park (eds), *British Social Attitudes, the 11th Report* (Aldershot: Dartmouth).

Worcester, R. (1994), 'Societal Values, Behaviour and Attitudes in Relation to the Human Dimensions of Global Environmental Change', paper presented at the 16th World Congress of the International Political Science Association, August, Berlin.

Wring, D. (1996), 'The Political Marketing Mix: A Framework for Election Campaign Analysis', a paper presented at the International Conference on Political Marketing, Judge Institute for Management Studies, University of Cambridge, 27–29 March.

Zald, M. N. (1987), 'The Future of Social Movements', in M. N. Zald and J. D. McCarthy (eds), *Social Movements in an Organizational Society* (New Brunswick, NJ: Transaction Books).

Zald, M. and Ash, R. (1966), 'Social Movement Organizations' *Social Forces*, 44: 327–41.

Zald, M. N. and McCarthy, J. D. (eds), (1987), *Social Movements in an Organizational Society* (New Brunswick, NJ: Transaction Books).

Index